THEORY OF SPINORS
An Introduction

An Introduction
Theory of Spinors

Moshe Carmeli
Ben Gurion University, Israel

Shimon Malin
Colgate University, USA

 **World Scientific**

NEW JERSEY • LONDON • SINGAPORE • BEIJING • SHANGHAI • HONG KONG • TAIPEI • CHENNAI

Published by

World Scientific Publishing Co. Pte. Ltd.
5 Toh Tuck Link, Singapore 596224
USA office: 27 Warren Street, Suite 401-402, Hackensack, NJ 07601
UK office: 57 Shelton Street, Covent Garden, London WC2H 9HE

British Library Cataloguing-in-Publication Data
A catalogue record for this book is available from the British Library.

First published 2000
Reprinted 2006

THEORY OF SPINORS: AN INTRODUCTION

Copyright © 2000 by World Scientific Publishing Co. Pte. Ltd.

All rights reserved. This book, or parts thereof, may not be reproduced in any form or by any means, electronic or mechanical, including photocopying, recording or any information storage and retrieval system now known or to be invented, without written permission from the Publisher.

For photocopying of material in this volume, please pay a copying fee through the Copyright Clearance Center, Inc., 222 Rosewood Drive, Danvers, MA 01923, USA. In this case permission to photocopy is not required from the publisher.

ISBN 981-02-4261-1
ISBN 981-256-472-1 (pbk)

Printed in Singapore by Utopia Press Pte Ltd

TO OUR GRANDCHILDREN

Nadav, Guy, Daniel, Shelly

and

Michaela

Preface

This is a textbook intended for advanced undergraduate and graduate students in physics and mathematics, as well as a reference for researchers. The book is based on lectures given during the years at the Ben Gurion University, Israel. Spinors are used extensively in physics; it is widely accepted that they are more fundamental than tensors and the easy way to see this fact is the results obtained in general relativity theory by using spinors, results that could not have been obtained by using tensor methods only. The book is written for the general physicist and not only to the workers in general relativity, even though the latter will find it most useful since it includes all what is needed in that theory.

But the foundations of the concept of spinors are groups; spinors appear as representations of groups. In this text we give a wide exposition to the relationship between the spinors and the representations of the groups. As is well known, both the spinors and the representations are widely used in the theory of elementary particles.

After presenting the origin of spinors from representation theory we, nevertheless, apply the theory of spinors to general relativity theory, and a part of the book is devoted to curved spacetime applications.

In the first four chapters we present the group-theoretical foundations of the concept of two-component spinors. Chapter 1 starts with an introduction to group theory emphasizing the rotation group. This followed by discussing representation theory in Chapter 2, including a brief outline of the infinite-dimensional case. Chapters 3 and 4 discuss in detail the Lorentz and the SL(2,C) groups. Here we give an extensive discussion on how two-component spinors emerge from the finite-dimensional representations of the group SL(2,C). Chapter 4 also includes the derivation of infinite-dimensional spinors as a generalization to the two-component spinors.

In Chapters 5 and 6 we apply the two-component spinors to a variety of problems in curved spacetime. In Chapter 5 we discuss the Maxwell, Dirac and Pauli spinors. Also given in this chapter the passage to the curved spacetime of spinors. The gravitational field spinors are subsequently discussed in detail in Chapter 6. Here we derive the curvature spinor and give the spinors equivalent to the Riemann, Weyl, Ricci and Einstein tensors.

In Chapter 7 we present the gauge field spinors and discuss their geometrical properties. As is well known, gauge fields are extremely important nowadays. The Euclidean gauge field spinors are finally discussed in Chapter 8.

All chapters of the book start with the ordinary physical material before introducing the spinors of that subject. Thus, for instance, the chapters dealing with the Lorentz group and gravitation start with detailed discussion of the theories of special relativity and general relativity.

It is a pleasure to thank our wifes Elisheva and Tova for creating the necessary atmosphere and for their patience while writing this book. We are grateful to the many students who attended the courses in spinors during the years for their suggestions which led to a better presentation of the material in the book. We also want to thank Silvia Behar for her help with the Index of the book. Finally, we want to thank Julia Goldbaum for the excellent job of typing the book, prepairing the Index, and for the many suggestions for improvements.

Moshe Carmeli
Beer Sheva, Israel

Shimon Malin
Hamilton, N.Y.
U.S.A.

Contents

Preface   vii

1 Introduction to Group Theory   1
  1.1 Review of Group Theory 1
      1.1.1 Group and Subgroup 1
      1.1.2 Normal Subgroup and Factor Group 3
      1.1.3 Isomorphism and Homomorphism 3
  1.2 The Pure Rotation Group SO(3) 4
      1.2.1 The Euler Angles 4
  1.3 The Special Unitary Group SU(2) 6
      1.3.1 Homomorphism between the Groups SO(3) and SU(2)   6
  1.4 Invariant Integrals over Groups 8
      1.4.1 Invariant Integral over the Group SO(3) 8
      1.4.2 Invariant Integral over the Group SU(2) 9
  1.5 Problems 10
  1.6 References for Further Reading 10

2 Representation Theory   11
  2.1 Some Basic Concepts 11
      2.1.1 Linear Operators 11
      2.1.2 Finite-Dimensional Representations 12
      2.1.3 Unitary Representations 13
  2.2 Representations of SO(3) and SU(2) 13
      2.2.1 Weyl's Method 13
      2.2.2 Infinitesimal Generators 14
      2.2.3 Basic Infinitisimal Operators 16
      2.2.4 Canonical Basis 16
      2.2.5 Unitary Matrices Corresponding to Rotations 17
  2.3 Matrix Elements of Representations 18
      2.3.1 The Spinor Representation of the Group SU(2) ... 19

		2.3.2	Matrix Elements of Representations	20
	2.4	2.3.3	Properties of $D_{mn}^j(u)$	21
			Differential Operators of Rotations	22
		2.4.1	Representation of SO(3) in Space of Functions	23
		2.4.2	The Differential Operators	24
		2.4.3	Angular Momentum Operators	26
	2.5		Infinite-Dimensional Representations	27
		2.5.1	Banach Space	27
		2.5.2	Hilbert Space	29
		2.5.3	Operators in a Banach Space	30
		2.5.4	General Definition of a Representation	30
		2.5.5	Continuous Representations	31
		2.5.6	Unitary Representations	31
	2.6		References for Further Reading	32

3 The Lorentz and SL(2,C) Groups — 35

	3.1		Elements of Special Relativity	35
		3.1.1	Postulates of Special Relativity	35
		3.1.2	The Galilean Transformation	37
		3.1.3	The Lorentz Transformation	38
		3.1.4	Derivation of the Lorentz Transformation	38
		3.1.5	The Cosmological Transformation	41
	3.2		The Lorentz Group	44
		3.2.1	Orthochronous Lorentz Transformation	46
		3.2.2	Subgroups of the Lorentz Group	46
	3.3		The Infinitesimal Approach	47
		3.3.1	Infinitesimal Lorentz Matrices	47
		3.3.2	Infinitesimal Operators	49
		3.3.3	Determination of the Representation by its Infinitesimal Operators	51
		3.3.4	Conclusions	52
		3.3.5	Unitarity Conditions	53
	3.4		The Group SL(2,C) and the Lorentz Group	54
		3.4.1	The Group SL(2,C)	54
		3.4.2	Homomorphism of the Group SL(2,C) on the Lorentz Group L	56
		3.4.3	Kernel of Homomorphism	58
		3.4.4	Subgroups of the Group SL(2,C)	59
		3.4.5	Connection with Lobachevskian Motions	60
	3.5		Problems	61

CONTENTS   xi

    3.6 References for Further Reading 62

4 Two-Component Spinors   **65**
    4.1 Spinor Representation of SL(2,C) 65
        4.1.1 The Space of Polynomials 65
        4.1.2 Realization of the Spinor Representation 66
        4.1.3 Two-Component Spinors 68
        4.1.4 Examples 71
    4.2 Operators of the Spinor Representation 73
        4.2.1 One-Parameter Subgroups 73
        4.2.2 Infinitesimal Operators 74
        4.2.3 Matrix Elements of the Spinor Operator $D(g)$ . . . 75
        4.2.4 Further Properties of Spinor Representations 77
    4.3 Infinite-Dimensional Spinors 77
        4.3.1 Principal Series of Representations 77
        4.3.2 Infinite-Dimensional Spinors 79
    4.4 Problems 81
    4.5 References for Further Reading 81

5 Maxwell, Dirac and Pauli Spinors   **83**
    5.1 Maxwell's Theory 83
        5.1.1 Maxwell's Equations in Curved Spacetime 86
    5.2 Spinors in Curved Spacetime 88
        5.2.1 Correspondence between Spinors and Tensors 88
        5.2.2 Raising and Lowering Spinor Indices 89
        5.2.3 Properties of the σ Matrices 89
        5.2.4 The Metric $g_{AB'CD'}$ and the Minkowskian Metric $\eta_{\mu\nu}$ 90
        5.2.5 Hermitian Spinors 91
    5.3 Covariant Derivative of a Spinor 92
        5.3.1 Spinor Affine Connections 92
        5.3.2 Spin Covariant Derivative 93
        5.3.3 A Useful Formula 95
    5.4 The Electromagnetic Field Spinors 95
        5.4.1 The Electromagnetic Potential Spinor 95
        5.4.2 The Electromagnetic Field Spinor 96
        5.4.3 Decomposition of the Electromagnetic Spinor 96
        5.4.4 Intrinsic Spin Structure 97
        5.4.5 Pauli, Dirac and Maxwell Equations 98
    5.5 Problems 100
    5.6 References for Further Reading 107

6 The Gravitational Field Spinors — 109
6.1 Elements of General Relativity — 109
6.1.1 Riemannian Geometry — 110
6.1.2 Principle of Equivalence — 118
6.1.3 Principle of General Covariance — 119
6.1.4 Gravitational Field Equations — 120
6.1.5 The Schwarzschild Solution — 123
6.1.6 Experimental Tests of General Relativity — 127
6.1.7 Equations of Motion — 133
6.1.8 Decomposition of the Riemann Tensor — 142
6.2 The Curvature Spinor — 143
6.2.1 Spinorial Ricci Identity — 144
6.2.2 Symmetry of the Curvature Spinor — 145
6.3 Relation to the Riemann Tensor — 146
6.3.1 Bianchi Identities — 147
6.4 The Gravitational Field Spinors — 148
6.4.1 Decomposition of the Riemann Tensor — 148
6.4.2 The Gravitational Spinor — 150
6.4.3 The Ricci Spinor — 152
6.4.4 The Weyl Spinor — 153
6.4.5 The Bianchi Identities — 157
6.5 Problems — 157
6.6 References for Further Reading — 163

7 The Gauge Field Spinors — 167
7.1 The Yang-Mills Theory — 167
7.1.1 Gauge Invariance — 167
7.1.2 Isotopic Spin — 168
7.1.3 Conservation of Isotopic Spin and Invariance — 168
7.1.4 Isotopic Spin and Gauge Fields — 169
7.1.5 Isotopic Gauge Transformation — 169
7.1.6 Field Equations — 171
7.2 Gauge Potential and Field Strength — 172
7.2.1 The Yang-Mills Spinor — 172
7.2.2 Energy-Momentum Spinor — 174
7.2.3 SU(2) Spinors — 176
7.2.4 Spinor Indices — 176
7.3 The Geometry of Gauge Fields — 177
7.3.1 Four-Index Tensor — 177
7.3.2 Spinor Formulation — 179

CONTENTS   xiii

	7.3.3 Comparison with the Gravitational Field	180
	7.3.4 Ricci and Einstein Spinors	182
7.4	References for Further Reading	183
8	**The Euclidean Gauge Field Spinors**	**185**
8.1	Euclidean Spacetime	185
	8.1.1 The Euclidean Dirac Equation	186
	8.1.2 Algebra of the Matrices s_μ	188
8.2	The Euclidean Gauge Field Spinors	190
	8.2.1 O(4) Two-Component Spinors	190
	8.2.2 Self-Dual and Anti-Self-Dual Fields	193
8.3	Problems	194
8.4	References for Further Reading	194

Index   **197**

Chapter 1

Introduction to Group Theory

In this chapter a brief discussion on group theory is given. This includes the concept of group and subgroup, normal subgroup and factor group. Isomorphism and homomorphism are subsequently discussed. This then followed by introducing the rotation group and the group SU(2), the aggregate of unitary matrices of order two and determinant unity. A homomorphism between the pure rotation group and the group SU(2) is subsequently established. The chapter is concluded with presenting invariant integrals over the groups.

1.1 Review of Group Theory

In this section the fundamental concepts of group theory are briefly presented. For details the reader is refered to the books of Pontrjagin, van der Waerden and others suggested at the end of the chapter.

1.1.1 Group and Subgroup

A non-empty set G of elements a, b, c,..., such as numbers, mappings, transformations, is called a *group* if the following *axioms* are satisfied:

(1) There exists an operation in the set G which associates to each two elements a and b of G a third element c of G. This operation is called

multiplication, and the element c is called the *product* of a and b, denoted by $c = ab$;

(2) The multiplication is *associative*, namely, if a, b and c are elements of G, then $(ab)\,c = a\,(bc)$;

(3) The set G contains a *right identity*, namely, there exists an element e such that $ae = a$ for each element a of G; and

(4) For each element a of G there exists a *right inverse* element, denoted by a^{-1}, such that $aa^{-1} = e$.

If the set G is finite, then the group G is called *finite* and the number of elements of G is called its *order*. Otherwise, the group G is called *infinite*. If the product of any two elements a and b of G is commutative, namely, $ab = ba$, the group is called *abelian*. In abelian groups the multiplication notation ab is replaced by an addition notation $a + b$, and the group operation is called *addition*. The identity is called *zero* and denoted by 0, and the inverse of a is called the *negative* of a and denoted by $-a$.

Since the product of group elements is associative, one writes for $(ab)\,c = a\,(bc)$ simply abc and for $(a+b)+c = a+(b+c)$ just $a+b+c$. The same holds for products of any number of elements. One can easily show (see Problem 1.1) that a right identity e is also a left identity, namely, $ea = a$, for any element a of G.

Likewise, a right inverse a^{-1} of a is also a left inverse, $a^{-1}a = e$. Hence the inverse of a^{-1} is simply a. Moreover, it follows that both the identity and the inverse are unique. This allows the use of the notation of the notation of algebra such as $a^{m+1} = a^m a$, with $a^1 = a$, for any natural number m. Negative powers of a are introduced by $a^{-m} = (a^{-1})^m$, $a^0 = e$. Hence $a^p a^q = a^{p+q}$, and $(a^p)^q = a^{pq}$, where p and q are integers.

An example of a group is the set of all nonzero rational numbers, if the rule of combination is ordinary multiplication. The identity is the number 1.

Another example of a group whose elements are not numbers is the aggregate of rotations of a plane or of space about a fixed point. Two rotations a and b are combined by performing the rotations successively. If b is carried out first and then a, the same result, i.e. the same final position of all points of the space, may also be obtained by a single rotation, denoted by ab. The group of rotations in space is an example of non-abelian group since it is not immaterial whether one performs first the rotation a and then b, or first b and then a. The identity of the rotation group is the identical transformation that leaves every point in its original position. The inverse of a rotation is the rotation in the opposite sense which cancels the first one.

1.1. REVIEW OF GROUP THEORY

A set H of elements of a group G is called a *subgroup* of G if it is a group with the same law of multiplication which operates in G. A necessary and sufficient condition for a subset H of a group G to be a subgroup is that if H contains two elements a and b it must also contain the element ab^{-1} (see Problem 1.2).

1.1.2 Normal Subgroup and Factor Group

Let G be a group and H a subgroup, and let a and b be two elements of G. One calles a and b *equivalent*, $a \propto b$, if ab^{-1} is an element of H. The group G is thus devided into classes of equivalent elements each called a *right coset* of H relative to G. It follows that if A is a right coset of H and a is an element of A then $A = Ha$. Moreover, every set of the form Hb is a right coset and the subgroup H itself is one of the cosets. One can also introduce *left cosets* of H, written in the form aH. They are obtained from an equivalence relation such that $a \propto b$ if $a^{-1}b$ belongs to H.

A subgroup N of a group G is called an *invariant* or *normal subgroup* of G if for every element n of N and a of G the element $a^{-1}na$ belongs to N. It follows that a necessary and sufficient condition for right and left cosets of a subgroup N to coincide is that N be a normal subgroup. Every group has at least two normal subgroups, the subgroup which includes only the identity, and the subgroup which coincides with the group itself. A group which has no normal subgroup except for these two subgroups is called *simple*.

If N is a normal subgroup of a group G and A and B are two cosets of N, $A = Na$, $B = Nb$, then AB is also a coset of N. The multiplication of cosets thus defined satisfies the group axioms, and the set of all cosets is called the *factor group* of G by the normal subgroup N and is denoted by G/N.

1.1.3 Isomorphism and Homomorphism

A mapping f of a group G on another group G' is called *isomorphism* if it (1) is one-to-one; and (2) preserves the multiplication. G and G' are then called *isomorphic*. The inverse f^{-1} of an isomorphism f is itself an isomorphism. An isomorphism of a group onto itself is called *automorphism*. The aggregate of all automorphisms of a group forms a group.

A mapping f of a group G on another group G' is called *homomorphism* if it preserves the operation of multiplication. The set N of all elements of G which go over into the identity of G' under the homomorphism is called the *kernel* of the homomorphism. If the kernel coincides with the identity of G then the homomorphism is an isomorphism. It follows that N is a normal of G, and G' is isomorphic to G/N. The isomorphism between G' and G/N is called the *natural isomorphism*.

The mapping f of a group G on G/N defined by associating with each element a of G the element $f(a) = A$ of G/N containing a is a homomorphism, called the *natural homomorphism* of a group on its factor group. If f is a homomorphism of a group G on another group G' and H is a (normal) subgroup of G, then $f(H)$ is a (normal) subgroup of G'. If f is a homomorphism of a group G on another group G', and g is a homomorphism of G' on a third group G'', then the mapping gf is a homomorphism of G on G''.

One finally notes that if f is a homomorphism of a group G on part of another group G' then the set of all elements of G' which are images of elements of G forms a subgroup of G'. Also, if $f^{-1}(H')$ is the set of all elements of G which go into $H' \subset G'$ under the homomorphism f, and if H' is a (normal) subgroup of the group G', then $f^{-1}(H')$ is also a (normal) subgroup of the group G.

1.2 The Pure Rotation Group SO(3)

A linear transformation g of the variables x_1, x_2, and x_3, which leaves the form $x_1^2 + x_2^2 + x_3^2$ invariant, is called a *three-dimensional rotation*. The aggregate of all such linear transformations g forms a continuous group, which is isomorphic to the set of all *real orthogonal* (namely, $gg^t = 1$, where g^t is the transposed of g) 3-dimensional matrices and is known as the *three-dimensional rotation group*. One can easily show that the determinant of every orthogonal matrix is equal to either $+1$, in which case the transformation describes *pure rotation*, or to -1, in which case it describes a *rotation-reflection*. The aggregate of all pure rotations forms a group, which is a subgroup of the 3-dimensional rotation group, and is known as the *pure rotation group*. We will be concerned with the 3-dimensional pure rotation group. This group is denoted by us by SO(3). (For more details, see in the sequel.)

1.2.1 The Euler Angles

Let g be an element of the group SO(3), i.e., a 3-dimensional orthogonal matrix with determinant unity. It is well known that one then can express each such element in terms of a set of three parameters. An example of

1.2. THE PURE ROTATION GROUP SO(3)

such parameters is that of the familiar Euler angles, which are defined as the three successive angles of rotation describing the transformation from a given Cartesian coordinate system to another one by means of three successive rotations performed in a specific sequence.

The sequence will be started by rotating the original system of axes **X** by an angle ϕ_1 clockwise about the z axis. The new coordinate system will be denoted by Ξ. One then has

$$\Xi = g(\phi_1)\,\mathbf{X}, \tag{1.1}$$

where the orthogonal matrix $g(\phi_1)$ is given by

$$g(\phi_1) = \begin{pmatrix} \cos\phi_1 & -\sin\phi_1 & 0 \\ \sin\phi_1 & \cos\phi_1 & 0 \\ 0 & 0 & 1 \end{pmatrix}. \tag{1.2}$$

We use the notation according to which $\mathbf{X} = (x,\,y,\,z) = (x_1,\,x_2,\,x_3)$, $\Xi = (\xi,\,\eta,\,\zeta)$, and $\mathbf{X}' = (x',\,y',\,z') = (x'_1,\,x'_2,\,x'_3)$, $\Xi' = (\xi',\,\eta',\,\zeta')$.

In the second stage the intermediate axes Ξ are rotated about its ξ axis clockwise by an angle θ to another intermediate set which is denoted Ξ', thus one has

$$\Xi' = g(\theta)\,\mathbf{X}', \tag{1.3}$$

where the orthogonal matrix $g(\theta)$ is given by

$$g(\theta) = \begin{pmatrix} 1 & 0 & 0 \\ 0 & \cos\theta & -\sin\theta \\ 0 & \sin\theta & \cos\theta \end{pmatrix}. \tag{1.4}$$

The ξ' axis is called the *line of nodes*. Finally the Ξ' axes are rotated clockwise by an angle ϕ_2 about the ξ' axis to produce the desired $\mathbf{X}'$ system of axes,

$$\mathbf{X}' = g(\phi_2)\,\Xi, \tag{1.5}$$

where the orthogonal matrix $g(\phi_2)$ is now given by

$$g(\phi_2) = \begin{pmatrix} \cos\phi_2 & -\sin\phi_2 & 0 \\ \sin\phi_2 & \cos\phi_2 & 0 \\ 0 & 0 & 1 \end{pmatrix}. \tag{1.6}$$

The matrix of the complete transformation

$$\mathbf{X}' = g\mathbf{X}, \tag{1.7}$$

is given, therefore, by the product of the matrices $g = g(\phi_2)\,g(\theta)\,g(\phi_1)$. Hence it is given by

$$g = \begin{pmatrix} \cos\phi_2 \cos\phi_1 & -\cos\phi_2 \sin\phi_1 & \sin\phi_2 \sin\theta \\ -\cos\theta \sin\phi_1 \sin\phi_2 & -\cos\theta \cos\phi_1 \sin\phi_2 & \\ \sin\phi_2 \cos\phi_1 & -\sin\phi_2 \sin\phi_1 & -\cos\phi_2 \sin\theta \\ +\cos\theta \sin\phi_1 \cos\phi_2 & +\cos\theta \cos\phi_1 \cos\phi_2 & \\ \sin\theta \sin\phi_1 & \sin\theta \cos\phi_1 & \cos\theta \end{pmatrix}. \quad (1.8)$$

The angles ϕ_1, θ, ϕ_2 are independent parameters, fully determining the rotation g. They are known as the *Euler angles*. By their definition, one has $0 \leq \phi_1 \leq 2\pi$, $0 \leq \theta \leq \pi$, and $0 \leq \phi_2 \leq 2\pi$ for the intervals of the angles ϕ_1, θ, and ϕ_2.

1.3 The Special Unitary Group SU(2)

Rotations can also be described by unitary matrices of order two and determinant unity. The aggregate of all such matrices provides a group which is usually denoted by SU(2). The relation between the groups SO(3) and SU(2) can be established as follows.

1.3.1 Homomorphism between the Groups SO(3) and SU(2)

Let x_l and x'_k, with $k, l = 1, 2, 3$, denote the coordinates of two Cartesian frames related by the transformation

$$x'_k = g_{kl} x_l, \quad (1.9)$$

where g_{kl} are elements of the matrix $g \in$ SO(3), and repeated indices means summation from 1 to 3. With each coordinate system x_k one associates a 2×2 Hermitian matrix P defined by

$$P = x_k \sigma^k = \begin{pmatrix} z & x+iy \\ x-iy & -z \end{pmatrix}, \quad (1.10)$$

where σ^k are the familiar Pauli spin matrices,

$$\sigma^1 = \begin{pmatrix} 0 & 1 \\ 1 & 0 \end{pmatrix}, \quad \sigma^2 = \begin{pmatrix} 0 & i \\ -i & 0 \end{pmatrix}, \quad \sigma^3 = \begin{pmatrix} 1 & 0 \\ 0 & -1 \end{pmatrix}. \quad (1.11)$$

1.3. THE SPECIAL UNITARY GROUP SU(2)

In terms of the matrix P one requires that the coordinates transform according to the formula

$$P' = uPu^\dagger, \tag{1.12}$$

where u is an element of the group SU(2),

$$P' = x'_l \sigma^l, \tag{1.13}$$

and $u^\dagger$ is the Hermitian conjugate of the matrix u. The relations between u of SU(2) and g of SO(3) are given by

$$g_{rs} = \frac{1}{2} \operatorname{Tr} \left(\sigma^r u \sigma^s u^\dagger \right), \tag{1.14}$$

$$u = \mp (1 + \sigma^r \sigma^s g_{rs})/2 (1 + \operatorname{Tr} g)^{1/2}, \tag{1.15}$$

where Tr stands for trace.

Accordingly, to each rotation g of the group SO(3) there correspond, by Eq. (1.15), two matrices $\mp u$ of the group SU(2) and, conversely, to each unitary matrix u of the group SU(2) there corresponds, by Eq. (1.14), some rotation g of the group SO(3). It thus follows that the group SU(2) is homomorphic to the group SO(3) (see Section 1.1). For example, the unitary matrices corresponding to the rotations $g(\phi_1)$, $g(\theta)$, and $g(\phi_2)$ given by Eqs. (1.2), (1.4) and (1.6) are easily found, using Eq. (1.15). They are given by

$$u(\phi_1) = \mp \begin{pmatrix} e^{i\phi_1/2} & 0 \\ 0 & e^{-i\phi_1/2} \end{pmatrix}, \tag{1.16a}$$

$$u(\theta) = \mp \begin{pmatrix} \cos\frac{\theta}{2} & i\sin\frac{\theta}{2} \\ i\sin\frac{\theta}{2} & \cos\frac{\theta}{2} \end{pmatrix}, \tag{1.16b}$$

and

$$u(\phi_2) = \mp \begin{pmatrix} e^{i\phi_2/2} & 0 \\ 0 & e^{-i\phi_2/2} \end{pmatrix}. \tag{1.16c}$$

A general rotation g, described by the matrix (1.8), will then correspond to the unitary matrix $u = u(\phi_2) u(\theta) u(\phi_1)$, and is thus given by

$$u = \mp \begin{pmatrix} \cos\frac{\theta}{2} e^{i(\phi_2+\phi_1)/2} & i\sin\frac{\theta}{2} e^{i(\phi_2-\phi_1)/2} \\ i\sin\frac{\theta}{2} e^{-i(\phi_2-\phi_1)/2} & \cos\frac{\theta}{2} e^{-i(\phi_2+\phi_1)/2} \end{pmatrix}. \tag{1.17}$$

1.4 Invariant Integrals over Groups

A function $y = f(g)$ is said to be defined over the group G if to each element g of G there corresponds a number y. If the group is taken to be the rotation group SO(3) and one uses the Euler angles as parameters then $f(g)$, where $g \in$ SO(3), becomes simply a function of the angles ϕ_1, θ, ϕ_2, i.e.,

$$f(g) = f(\phi_1, \theta, \phi_2). \tag{1.18}$$

The function f then satisfies

$$f(\phi_1 + 2\pi, \theta, \phi_2) = f(\phi_1, \theta, \phi_2), \tag{1.19a}$$

$$f(\phi_1, \theta, \phi_2 + 2\pi) = f(\phi_1, \theta, \phi_2). \tag{1.19b}$$

1.4.1 Invariant Integral over the Group SO(3)

The integral $\int f(g)\, dg$ is then called *invariant integral* of the function $f(g)$ over the group SO(3) if it satisfies

$$\int f(gg_0)\, dg = \int f(g_0 g)\, dg = \int f(g)\, dg \tag{1.20}$$

for any $g_0 \in$ SO(3), and

$$\int f(g^{-1})\, dg = \int f(g)\, dg. \tag{1.21}$$

The expression dg is called a *measure*. When the Euler angles are used to parametrize the elements g of the group SO(3), one can write dg in terms of the angles ϕ_1, θ, ϕ_2 as

$$dg = \frac{1}{8\pi^2} \sin\theta\, d\phi_1\, d\theta\, d\phi_2. \tag{1.22}$$

One then can easily verify that it satisfies

$$\int dg = 1. \tag{1.23}$$

The integration limits extend over the whole domain of definitions of the variables, i.e., $0 \leq \phi_1 \leq 2\pi$, $0 \leq \theta \leq \pi$, and $0 \leq \phi_2 \leq 2\pi$.

1.4.2 Invariant Integral over the Group SU(2)

The concepts of functions defined over the group SO(3) and invariant integrals defined over the rotation group SO(3) can easily be extended to the special unitary group SU(2). Again, a function $f(u)$ defined over the group SU(2) can be considered as a function of the angles ϕ_1, θ, ϕ_2, i.e.,

$$f(u) = f(\phi_1, \theta, \phi_2) \tag{1.24}$$

if the Euler angles are used for parametrization. The analogous periodicity conditions to those of Eq. (1.19) for functions defined over SO(3) will now be

$$f(\phi_1 + 4\pi, \theta, \phi_2) = f(\phi_1, \theta, \phi_2), \tag{1.25a}$$

$$f(\phi_1, \theta, \phi_2 + 4\pi) = f(\phi_1, \theta, \phi_2), \tag{1.25b}$$

$$f(\phi_1 + 2\pi, \theta, \phi_2 + 2\pi) = f(\phi_1, \theta, \phi_2). \tag{1.25c}$$

The invariant integral over the group SU(2) then satisfies

$$\int f(uu_0)\, du = \int f(u_0 u)\, du = \int f(u)\, du \tag{1.26}$$

for any $u \in \mathrm{SU}(2)$, and

$$\int f(u^{-1})\, du = \int f(u)\, du. \tag{1.27}$$

The measure du can then be expressed in terms of the Euler angles as

$$du = \frac{1}{16\pi^2} \sin\theta\, d\phi_1\, d\theta\, d\phi_2. \tag{1.28}$$

It can be shown that it satisfies

$$\int du = 1. \tag{1.29}$$

The integration limits here will be: $0 \leq \phi_1 \leq 4\pi$, $0 \leq \theta \leq \pi$, and $0 \leq \phi_2 \leq 2\pi$.

In the next chapter the theory of representations of groups is given and applied to the rotation group.

1.5 Problems

1.1. Show that a right identity e is also a left identity, namely, $ea = a$, for any element a of a group G. Show also that a right inverse a^{-1} of a is also a left inverse, namely, $a^{-1}a = e$.

Solution: The solution is left for the reader.

1.2. Show that a necessary and sufficient condition for a subset H of a group G to be a subgroup is that if H contains two elements a and b it must also contain the element ab^{-1}.

Solution: The solution is left for the reader.

1.6 References for Further Reading

M. Carmeli and S. Malin, *Representations of the Rotation and Lorentz Groups* (Marcel Dekker, New York and Basel, 1976).

C. Chevalley, *Theory of Lie Groups* (Princeton University Press, New Jersey, 1962). (Section 1.1)

L.P. Eisenhart, *Continuous Groups of Transformations* (Dover Publications, Inc., New York, 1961). (Section 1.1)

H. Goldstein, *Classical Mechanics* (Addison-Wesley Publishing Co., Reading, Mass., 1965). (Section 1.3)

M.A. Naimark, *Linear Representations of the Lorentz Group* (Pergamon Press, New York, 1964). (Sections 1.3, 1.4)

L. Pontrjagin, *Topological Groups* (Princeton University Press, Princeton, New Jersey, 1946). (Section 1.1)

A. Salam, *The formalism of Lie groups, Lecture Notes*, 1960.

B.L. van der Waerden, *Modern Algebra* (Fredric Ungar Publishing Co., New York, 1953). (Section 1.1)

A. Weil, *Actualites Sci. Ind.*, No. 869 (1938); *L'integration dans les groups topologiques et ces applications* (Hermann et Cie., Paris, 1940). (Section 1.4)

Chapter 2

Representation Theory

In the last chapter the important concept of groups was discussed. In this chapter the theory of representations of groups is given and applied to the rotation group and the group SU(2). The spinor representation of the group SU(2), along with the matrix elements, are then given. This subsequently followed by finding the differential operators of the rotations. The more complicated theory of infinite-dimensional representations is briefly given in the last section of the chapter.

2.1 Some Basic Concepts

In this section the fundamentals of the theory of finite-dimensional representations are given. For more details the reader is reffered to the books of Naimark, of Gelfand, Graev, and Vilenkin, and of others given in the suggested references at the end of the chapter. The more complicated theory of infinite-dimensional representations is also given in the last section of the chapter.

2.1.1 Linear Operators

Let S be a linear space and x a vector in it. A function $A(x)$ is called an *operator* in S if for any vector x of S there corresponds a vector $y = A(x)$ of S. An operator A in S is then called *linear* if $A(x+y) = A(x) + A(y)$ and $A(\alpha x) = \alpha A(x)$, for any x, y of S and a complex number α. Addition of two operators A and B is defined in the space S by $(A+B)x = Ax + Bx$ for all vectors x of S. Similarly, multiplication by a number α and multiplication of operators A and B in the space S are defined by $(\alpha A) x = \alpha (Ax)$ and $(AB) x = A(Bx)$. If, furthermore, A and B are linear operators in S then $A + B$, αA, and AB are also linear operators in S.

Linear operators in a finite-dimensional space S can be represented as matrices by introducing a basis, $e_1, ..., e_n$, in S. Accordingly, if A is a linear operator in the space S, then Ae_k can be written as a linear combination of $e_1, ..., e_n$, or,

$$Ae_k = \sum_{j=1}^{n} A_{jk} e_j; \quad (k = 1, \ldots, n). \tag{2.1}$$

A_{jk} are the elements of the matrix representing the operator A relative to the basis $e_1, ..., e_n$. One can show that the operator A is completely determined by its matrix A_{ij}. Furthermore, the operations of addition, multiplication by a number, and multiplication of operators correspond to the same operations of their matrices relative to a fixed basis.

2.1.2 Finite-Dimensional Representations

Let G be a group and g an arbitrary element of G. A correspondence $g \to D(g)$ of each element g of the group G to a linear operator $D(g)$ in a finite-dimensional space S is called a *representation* if: (1) $D(g_1) D(g_2) = D(g_1 g_2)$ and (2) $D(e)$ is the unit element in S, where e is the identity element of G. The space S is called the *space of representation* and its dimension is called the *dimension of the representation*. (For the more general definition of a representation see Subsection 2.5.4.)

Two finite-dimensional representations $g \to D_1(g)$ and $g \to D_2(g)$ of the group G in two spaces S_1 and S_2 having the same dimensions, respectively, are called *equivalent* if bases in the spaces S_1 and S_2 can be chosen so that the matrices of the operators $D_1(g)$ and $D_2(g)$ are identical. A subspace S' of the space S is called *invariant* with respect to the representation $g \to D(g)$ if for every vector x of S' one finds that $D(g) x$ is also a vector in S' for all elements g of the group G. If there are no invariant subspaces in the space S with respect to the representation $g \to D(g)$, except for the trivial cases of the null subspace and the whole space, the representation is then called *irreducible*.

A representation $g \to D(g)$ of a group G is called *continuous* if $D(g)$ is a continuous operator function on the group G. (An operator function $D(g)$ is called continuous on a group G if the elements of the matrix of $D(g)$, relative to a fixed basis, are continuous functions on G. This definition of continuity of $D(g)$ does not depend on the choice of the basis since the matrix elements relative to another basis are linear combinations, with constant coefficients, of the matrix elements relative to the original basis.) Only continuous representations will be considered here.

2.1.3 Unitary Representations

A linear space is called *Euclidean* if from each two vectors x and y of it one can define a function, called the *scalar product* of x and y, denoted by (x, y), which satisfies:

(1) $(x, x) \geq 0$, $(x, x) = 0$ if and only if $x = 0$;
(2) $(y, x) = \overline{(x, y)}$;
(3) $(\alpha x, y) = \alpha (x, y)$;
(4) $(x_1 + x_2, y) = (x_1, y) + (x_2, y)$.

One can show that a scalar product can be introduced in every finite-dimensional space. (The infinite-dimensional case is discussed in the appendix at the end of the chapter.)

An operator D in a finite-dimensional Euclidean space E is called *unitary* if it preserves the scalar product, namely, $(Dx, Dy) = (x, y)$ for all x, y of the space E. A representation $g \to D(g)$ is called *unitary* if all its operators $D(g)$ are unitary.

In the following we find the irreducible representations of the three-dimensional pure rotation group. This is done by Weyl's method which makes use of the homomorphism of the special unitary group of order two onto the rotation group. The representations are expressed in terms of the angle of rotation in a specified direction and the spherical angles of the direction of the rotation.

2.2 Representations of SO(3) and SU(2)

We have seen that the unimodular unitary group of order two, SU(2), is homomorphic to the pure rotation group SO(3) such that to every rotation g of SO(3) there correspond two matrices $+u$ and $-u$ of SU(2) and, conversely, to every element u of SU(2) there corresponds some rotation g of SO(3).

2.2.1 Weyl's Method

It thus follows that the description of the representations (see Section 2.1) of the group SO(3) is equivalent to that of the group SU(2); a representation $g \to D(g)$ of the group SO(3) is single- or double-valued according to whether or not $D(u)$ is equal to $D(-u)$. The use of the group SU(2) for finding the representations of the group SO(3) was originally suggested by H. Weyl and has been wildly adopted when the Euler angles are used to parametrize the groups. The advantage of Weyl's method is in giving the *double-valued* representations along with the proper representations. The double-valued representations are important in physical problems dealing with spin-like properties of particles whose spins are half integers.

We point out that, by using Weyl's method, one can obtain a general invariant result that is a function of the element $u \in$ SU(2), valid for any parametrization one uses to describe the rotation. To find the representations of the group SO(3) in terms of a specific set of parameters, one has merely to express u in terms of these parameters, as is the case when the Euler angles are adopted.

In addition, by having the results as functions over the group SU(2), certain relations will be obtained which are invariant under change of the parameters. As an example, the orthogonality relations between the matrix elements of the irreducible representation can be written in the form of an invariant integral over the group SU(2). Hence the relations are valid for any parametrization.

2.2.2 Infinitesimal Generators

An orthogonal matrix describing a rotation with an angle ψ about some direction

$$\mathbf{n} = (\sin\theta\cos\phi,\ \sin\theta\sin\phi,\ \cos\theta) \tag{2.2}$$

is given by

$$g_{rs} = \delta_{rs}\cos\psi + n_r n_s (1 - \cos\psi) - \epsilon_{rst} n^t \sin\psi, \tag{2.3}$$

where r, s and t take the values from 1 to 3. Rotations $g_1(\psi)$, $g_2(\psi)$ and $g_3(\psi)$ around Ox_1, Ox_2 and Ox_3 axes are then obtained from Eq. (2.3) by putting the proper values for the polar angles θ and ϕ. These matrices are given by

$$g_1(\psi) = \begin{pmatrix} 1 & 0 & 0 \\ 0 & \cos\psi & -\sin\psi \\ 0 & \sin\psi & \cos\psi \end{pmatrix}, \tag{2.4a}$$

$$g_2(\psi) = \begin{pmatrix} \cos\psi & 0 & \sin\psi \\ 0 & 1 & 0 \\ -\sin\psi & 0 & \cos\psi \end{pmatrix}, \tag{2.4b}$$

2.2. REPRESENTATIONS OF SO(3) AND SU(2)

$$g_3(\psi) = \begin{pmatrix} \cos\psi & -\sin\psi & 0 \\ \sin\psi & \cos\psi & 0 \\ 0 & 0 & 1 \end{pmatrix}. \tag{2.4c}$$

Infinitesimal Matrices

The *infinitesimal matrices* g_r, corresponding to rotations about the axis Ox_r are defined by

$$g_r = \left[\frac{dg_r(\psi)}{d\psi}\right]_{\psi=0}, \tag{2.5}$$

and satisfy the *commutation relations*

$$[g_r, g_s] = \epsilon_{rst} g_t, \tag{2.6}$$

where $[g_r, g_s] = g_r g_s - g_s g_r$.

The matrices g_r's are given by

$$g_1 = \begin{pmatrix} 0 & 0 & 0 \\ 0 & 0 & -1 \\ 0 & 1 & 0 \end{pmatrix}, \tag{2.7a}$$

$$g_2 = \begin{pmatrix} 0 & 0 & 1 \\ 0 & 0 & 0 \\ -1 & 0 & 0 \end{pmatrix}, \tag{2.7b}$$

$$g_3 = \begin{pmatrix} 0 & -1 & 0 \\ 1 & 0 & 0 \\ 0 & 0 & 0 \end{pmatrix}, \tag{2.7c}$$

and one has the relation

$$g_r(\psi) = \exp(\psi g_r). \tag{2.8}$$

Let us denote a representation of the group SO(3) in an n-dimensional Euclidean space R by $g \to D(g)$ and, for convenience, we denote

$$A_r(\psi) = D(g_r)(\psi). \tag{2.9}$$

$A_r(\psi)$ are called the *basic one-parameter groups* of the given representation and define one-parameter groups of operators that satisfy $A_r(\psi_1) A_r(\psi_2) = A_r(\psi_1 + \psi_2)$; they are differentiable functions of ψ and may be expanded as $A_r(\psi) = \exp(\psi A_r)$, where A_r is defined by Eq. (2.10)

2.2.3 Basic Infinitisimal Operators

The *basic infinitesimal operators* of the representation are then obtained by

$$A_r = \left[\frac{dA_r(\psi)}{d\psi}\right]_{\psi=0}. \tag{2.10}$$

A representation of the group SO(3) is *uniquely determined* by its basic infinitesimal operators A_r. The determination of all the finite-dimensional representations of the group SO(3) is based on the fact that the operators A_r satisfy the *same* commutation relations that exist among the infinitesimal matrices g_r:

$$[A_r, A_s] = \epsilon_{rst} A_t. \tag{2.11}$$

The operators A_r are skew-Hermitian, $A^\dagger = -A_r$, since, without loss of generality, every finite-dimensional representation of SO(3) can be considered to be unitary. (An operator B in a finite-dimensional Euclidean space E is called *adjoint* to the operator A in the same space if $(Ax, y) = (x, By)$ for all x, y of E. The adjoint of an operator A is usually denoted by $A^\dagger$. It can be shown that for any linear operator A there exists one and only one adjoint operator $A^\dagger$, and that the adjoint operator to $A^\dagger$ is A. An operator A is called *Hermitian* if $A^\dagger = A$. An operator A can be shown to be unitary if and only if $A^\dagger A = 1$.)

2.2.4 Canonical Basis

Defining the new operators

$$L_\mp = iA_1 \pm A_2, \tag{2.12a}$$

$$L_3 = iA_3, \tag{2.12b}$$

one then finds for the commutation relations of the infinitesimal generators L_+, L_-, and L_3 the following:

$$[L_\mp, L_3] = \pm L_\mp, \tag{2.13a}$$

$$[L_+, L_-] = 2L_3, \tag{2.13b}$$

$$L_+^\dagger = L_-, \quad L_3^\dagger = L_3. \tag{2.13c}$$

The problem of determining the representation is then reduced to the determination of the operators $L_\mp$ and L_3 satisfying the conditions (2.13).

2.2. REPRESENTATIONS OF SO(3) AND SU(2)

This problem is solved by the following: every finite-dimensional representation of the group SO(3) is uniquely determined by a non-negative integer or half-integer j, the *weight* of the representation.

The space of the representation corresponding to such a number j has the dimension $2j+1$; the operators $L_\mp$ and L_3 of the representation are given relative to its *canonical basis* $f_{-j}, f_{-j+1}, ..., f_j$ by

$$L_\pm f_m = [(j \mp m)(j \pm m + 1)]^{1/2} f_{m\pm 1}, \qquad (2.14a)$$

$$L_3 f_m = m f_m, \qquad (2.14b)$$

where $m = -j, -j+1, ..., j$.

It also follows that for each j there corresponds an irreducible representation of SO(3). If the operators $L_\mp$ and L_3 of a representation of SO(3) in a $(2j+1)$-dimensional space are given relative to some basis $f_{-j}, f_{-j+1}, ..., f_j$, then by Eqs. (2.14) that representation is irreducible.

2.2.5 Unitary Matrices Corresponding to Rotations

We now find the unitary matrix u corresponding to the rotation g, of Eq. (2.3). The matrices u and g are related by Eqs. (1.14) and (1.15). A direct calculation then gives:

$$u = \mp \begin{pmatrix} \cos\frac{\psi}{2} + i\sin\frac{\psi}{2}\cos\theta & i\sin\frac{\psi}{2}\sin\theta e^{i\phi} \\ i\sin\frac{\psi}{2}\sin\theta e^{-i\phi} & \cos\frac{\psi}{2} - i\sin\frac{\psi}{2}\cos\theta \end{pmatrix}. \qquad (2.15)$$

This is the unitary matrix $u \in SU(2)$ corresponding to a rotation with an angle ψ around the direction $\mathbf{n}$ specified by the spherical angles θ and ϕ. The corresponding matrix, when the Euler angles are employed, was given in Eq. (1.17). It will be noted that

$$u(-\psi, \theta, \phi) = u^{-1}(\psi, \theta, \phi). \qquad (2.16)$$

The unitary matrices $u_1(\psi)$, $u_2(\psi)$ and $u_3(\psi)$ corresponding to the rotations $g_1(\psi)$, $g_2(\psi)$ and $g_3(\psi)$ around the axes of coordinates Ox_1, Ox_2 and Ox_3, can be obtained from Eq. (2.15) by putting the appropriate values for the angles θ and ϕ. They are:

$$u_1(\psi) = \mp \begin{pmatrix} \cos\frac{\psi}{2} & i\sin\frac{\psi}{2} \\ i\sin\frac{\psi}{2} & \cos\frac{\psi}{2} \end{pmatrix}, \qquad (2.17a)$$

$$u_2(\psi) = \mp \begin{pmatrix} \cos\dfrac{\psi}{2} & -\sin\dfrac{\psi}{2} \\ \sin\dfrac{\psi}{2} & \cos\dfrac{\psi}{2} \end{pmatrix}, \tag{2.17b}$$

$$u_3(\psi) = \mp \begin{pmatrix} e^{i\psi/2} & 0 \\ 0 & e^{-i\psi/2} \end{pmatrix}. \tag{2.17c}$$

Using these matrices, the operators $A_r(\psi)$ of the group SU(2) will be determined in the next chapter.

The infinitesimal matrices u_r, corresponding to rotations around Ox_r, are given by

$$u_r = \left[\frac{du_r(\psi)}{d\psi}\right]_{\psi=0}, \tag{2.18}$$

and explicitly,

$$u_1 = \mp\frac{1}{2}\begin{pmatrix} 0 & i \\ i & 0 \end{pmatrix}, \quad u_2 = \mp\frac{1}{2}\begin{pmatrix} 0 & -1 \\ 1 & 0 \end{pmatrix}, \quad u_3 = \mp\frac{1}{2}\begin{pmatrix} i & 0 \\ 0 & -i \end{pmatrix}. \tag{2.19}$$

These are related to the Pauli matrices, Eq. (1.11), by

$$u_r = \mp\frac{i}{2}\sigma^r. \tag{2.20}$$

2.3 Matrix Elements of Representations

A matrix u of the group SU(2) can be considered as that of a linear transformation of the space of all pairs of complex numbers (ξ^1, ξ^2):

$$\xi'^p = \sum_{q=1}^{2} u_{pq}\xi^q \quad (p=1,2). \tag{2.21}$$

A representation of the group SU(2) can be obtained if one considers several pairs $(\xi_1^1, \xi_1^2), \ldots, (\xi_k^1, \xi_k^2)$ and forms all products $\xi_1^{p_1}\ldots\xi_k^{p_k}$, letting $p_1, \ldots, p_k$ take the values 1, 2, independently. Under the transformation (2.21), this product transforms like

$$\xi'^{p_1}_1\ldots\xi'^{p_k}_k = \sum_{q_1,\ldots,q_k=1}^{2} u_{p_1 q_1}\ldots u_{p_k q_k}\xi_1^{q_1}\ldots\xi_k^{q_k}. \tag{2.22}$$

2.3. MATRIX ELEMENTS OF REPRESENTATIONS

The product $\xi_1^{p_1}...\xi_k^{p_k}$ may be considered as a vector in the linear space R_k of all 2^k complex numbers $\xi^{p_1\cdots p_k}$. The linear transformation $D^{(k)}(u)$ of the space R_k is then given by

$$\xi'^{p_1\cdots p_k} = \sum_{q_1,\ldots,q_k=1}^{2} u_{p_1 q_1}\cdots u_{p_k q_k}\xi^{q_1\cdots q_k}. \tag{2.23}$$

2.3.1 The Spinor Representation of the Group SU(2)

The correspondence $u \to D^{(k)}(u)$ is a representation of the group SU(2), *not* irreducible in general, since the subspace S_k of R_k of all symmetrical vectors ξ is invariant with respect to all the operators $D^{(k)}(u)$. The correspondence $u \to D^{(k)}(u)$ is irreducible, however, in the subspace S_k. We denote this representation by Z_k. It is called the *spinor representation* of the group SU(2) and is of weight $k/2$.

An equivalent realization of the representation Z_k is obtained if one identifies the subspace S_k with the $(k+1)$-dimensional space of homogeneous polynomials $p(z_1, z_2)$ of degree k in the two complex variables z_1 and z_2 and sets up a one-to-one correspondence between ξ of S_k and $p(z_1, z_2)$ in the form

$$p(z_1, z_2) = \sum_{p_1,\ldots,p_k=1}^{2} \xi^{p_1\cdots p_k} z_{p_1}\cdots z_{p_k}. \tag{2.24}$$

The operator $D^{(k)}(u)$ for this new realization in the space of polynomials S_k is then given by

$$D^{(k)}(u) p(z_1, z_2) = p(z_1', z_2'), \tag{2.25a}$$

where

$$z_q' = \sum_{p=1}^{2} u_{pq} z_p \quad (q = 1, 2). \tag{2.25b}$$

Introducing now a new variable $z = z_1/z_2$, the polynomial $p(z_1, z_2)$ can then be written as $z_2^k p(z)$, where $p(z)$ is a polynomial in the variable z of degree not exceeding k. The operators $D^{(k)}(u)$ of the representation z_k are, accordingly, given by

$$D^{(k)}(u) p(z) = (u_{12} z + u_{22})^k p\left(\frac{u_{11} z + u_{21}}{u_{12} z + u_{22}}\right). \tag{2.26}$$

This equation gives, in particular, the operators $A_r(\psi) = D(u_r(\psi))$ when the matrices $u_r(\psi)$, Eqs. (2.17), are used. (For the determination of the operators $A_r(\psi)$, one needs $u_r(\psi)$ only for small values of ψ. The signs in Eqs. (2.17) are determined by the conditions $\lim u_r(\psi) = 1$ when $\psi \to 0$; hence the + sign must be used.)

2.3.2 Matrix Elements of Representations

It follows that every irreducible finite-dimensional representation of the group SU(2) is uniquely determined by some non-negative integer or half-integer $j = k/2$, the *weight* of the representation. Conversely, for any non-negative integer or half-integer j, there exists an irreducible representation of the group SU(2) of weight j. A representation of weight j can be realized as the spinor representation Z_k, where $k = 2j$; and every finite-dimensional irreducible representation of the group SU(2) is equivalent to one of the representations Z_k.

The functions

$$f_m(z) = \frac{(-z)^{j-m}}{[(j-m)!(j+m)!]^{1/2}}, \qquad (2.27)$$

where $m = -j, -j+1, ..., j$ then form a canonical basis for the representation z_k in the space S_k. Using Eq. (2.26), one finds

$$D^{(k)}(u) f_n(z) = \sum_{m=-j}^{j} D^j_{mn}(u) f_m(z), \qquad (2.28)$$

where $D^j_{mn}(u)$ are the matrix elements of the operator $D(u)$ of the irreducible representation of weight j relative to the canonical basis, which corresponds to an arbitrary rotation g. Its explicit expression is

$$D^j_{mn}(u) = (-1)^{2j-m-n} \left[\frac{(j-m)!(j+m)!}{(j-n)!(j+n)!}\right]^{1/2}$$
$$\times \sum \binom{j-n}{a}\binom{j+n}{j-m-a} u_{11}^a u_{12}^{j-m-a} u_{21}^{j-n-a} u_{22}^{m+n+a}, \qquad (2.29)$$

where a runs from $a = \max(0, -m, -n)$ to $a = \min(j-m, j-n)$, and

$$\binom{m}{n} = \frac{m!}{(m-n)!n!}. \qquad (2.30)$$

In Eq. (2.29) the indices m and n take the values $-j, -j+1, ..., j$ and $j = 0, 1/2, 1, 3/2, 2, ...$.

2.3. MATRIX ELEMENTS OF REPRESENTATIONS

It will be noted that $D^j_{mn}(-u) = (-1)^{2j} D^j_{mn}(u)$. Thus the representation is single-valued for integer j and double-valued for half-integer j. In the sequel the matrix u of Eq. (2.15) will be taken with the + sign.

To find the matrix elements (2.29) in terms of the variables ψ, θ and ϕ we simply substitute for u_{pq} their expressions as functions of these variables as given by Eq. (2.15). (One can easily find the expression of D^j in terms of Euler's angles.) One obtains

$$D^j_{mn}(\psi, \theta, \phi) = (-1)^{2j-m-n} \left[\frac{(j-m)!(j+m)!}{(j-n)!(j+n)!}\right]^{1/2}$$

$$\times \left(i\sin\frac{\psi}{2}\sin\theta e^{-i\phi}\right)^{m-n} \left(\cos\frac{\psi}{2} - i\sin\frac{\psi}{2}\cos\theta\right)^{m+n} S(j, m, n; x). \quad (2.31)$$

Here we have used the notation

$$S(j, m, n; x) = 2^{m-j}(j-n)!(j+n)!$$

$$\times \sum \frac{(x+1)^a (x-1)^{j-m-a}}{a!(j-n-a)!(j-m-a)!(a+m+n)!}, \quad (2.32)$$

where x is defined by

$$x = 1 - 2\sin^2\frac{\psi}{2}\sin^2\theta. \quad (2.33)$$

It will be noted that the function $S(j, m, n; x)$ is equal to the Jacobi polynomial $p_s^{\alpha\beta}(x)$ when $s = j - \frac{1}{2}(|m+n| + |m-n|)$, $\alpha = |m-n|$, and $\beta = |m+n|$.

2.3.3 Properties of $D^j_{mn}(u)$

Finally we discuss the properties of the matrices $D^j_{mn}(u)$.

One first notices that the matrices $D^j(u)$ are unitary. The correspondence $u \to D^j(u)$ is a representation of the group SU(2). Therefore one has

$$D^j_{mn}(u_1 u_2) = \sum_{n'=-j}^{j} D^j_{mn'}(u_1) D^j_{n'n}(u_2). \quad (2.34)$$

Furthermore, one has

$$D^j(u^{-1}) = [D^j(u)]^{-1} = [D^j(u)]^\dagger, \quad (2.35a)$$

or

$$D^j_{mn}(u)^{-1} = \overline{D}^j_{nm}(u). \qquad (2.35b)$$

Denoting now by γ the unitary matrix

$$\gamma = \begin{pmatrix} e^{-i\psi/2} & 0 \\ 0 & e^{i\psi/2} \end{pmatrix}, \qquad (2.36)$$

where ψ is a real number. If one applies now the representation formula (2.26), where $p(z)$ is taken as the basis functions $f_m(z)$ of Eq. (2.27), one obtains

$$D_\gamma f_m(z) = (-1)^{j-m} e^{ij\psi} \frac{\left(e^{-i\psi}z\right)^{j-m}}{\sqrt{(j-m)!(j+m)!}} = e^{im\psi} f_m(z). \qquad (2.37)$$

Hence the matrix $D^j(\gamma)$ is diagonal, and $D^j_{nn}(\gamma) = e^{in\psi}$. Furthermore, one easily finds that

$$D^j_{mn}(\gamma u) = e^{im\psi} D^j_{mn}(u), \qquad (2.38a)$$

$$D^j_{mn}(u\gamma) = e^{in\psi} D^j_{mn}(u). \qquad (2.38b)$$

We conclude this section by giving the orthogonality relation that the matrices D^j satisfy:

$$\int D^{j_1}_{m_1 n_1}(u) \overline{D}^{j_2}_{m_2 n_2}(u)\, du = (2j_1+1)^{-1} \delta_{j_1 j_2} \delta_{m_1 m_2} \delta_{n_1 n_2}. \qquad (2.39)$$

Relations similar to (2.39) are valid for any compact group. See, for example, the book of Pontrjagin.

2.4 Differential Operators of Rotations

We are now in a position to find the differential operators corresponding to infinitesimal rotations about the coordinate axis, namely, the operators A_1, A_2 and A_3 and, consequently, the operators $L_\mp$ and L_3. These operators are well known in the literature when the Euler angles are employed. We here derive these operators in terms of the variables ψ, θ and ϕ.

2.4.1 Representation of SO(3) in Space of Functions

Let $g \to D(g)$ be an irreducible representation of weight j of the group SO(3) and let $D_{mn} = D_{mn}^j$ be its matrix elements. We consider these elements as functions of the rotation g, $D_{mn} = D_{mn}(g)$. Since $g \to D(g)$ is a representation, one has

$$D(gg') = D(g) D(g'). \tag{2.40}$$

In terms of matrix elements, the last relation is

$$D_{mn}(gg') = \sum_{q=-j}^{j} D_{mq}(g) D_{qn}(g'), \tag{2.41}$$

where $D_{mn}(gg')$ are the matrix elements of the operators $D(gg')$.
Define now a transformation U such that

$$U(g') D_{mn}(g) = D_{mn}(gg'). \tag{2.42}$$

Comparing Eqs. (2.41) and (2.42) we obtain

$$U(g') D_{mn}(g) = \sum_{q=-j}^{j} D_{qn}(g') D_{mq}(g). \tag{2.43}$$

Furthermore, one can show that

$$U(g') U(g'') = U(g'g''). \tag{2.44}$$

It thus follows that the transformation $U(g')$ realizes a representation of the group SO(3) in the space of $2j+1$ functions of the mth row of the matrix $D(g)$ [compare Eq. (2.28)], and that the matrix elements of $U(g')$ are $D_{qn}(g')$.

The representation $g' \to U(g')$ in the space of functions $D_{mq}(g)$, $q = -j, -j+1, ..., j$, is irreducible, and the $D_{mq}(g)$ form a canonical basis in this space. Hence the operators $L_{\mp}$ and L_3 of this representation satisfy the relation (2.14), i.e.,

$$L_{\pm} D_{mn}^j(g) = [(j \pm n + 1)(j \mp n)]^{1/2} D_{m,n\pm 1}^j(g), \tag{2.45a}$$

$$L_3 D_{mn}^j(g) = n D_{mn}^j(g). \tag{2.45b}$$

2.4.2 The Differential Operators

To find the operators A_r we take g' as the rotation through some angle α around the axis Ox_r and expand the relation (2.42) in powers of α. Expansion of $D_{mn}(gg')$, which we denote by $D_{mn}\left(\tilde{\psi}, \tilde{\theta}, \tilde{\phi}\right)$, gives

$$D_{mn}\left(\tilde{\psi}, \tilde{\theta}, \tilde{\phi}\right) = D_{mn}(\psi, \theta, \phi)$$

$$+\alpha \left[\frac{\partial D_{mn}}{\partial \psi}\frac{d\tilde{\psi}}{d\alpha} + \frac{\partial D_{mn}}{\partial \theta}\frac{d\tilde{\theta}}{d\alpha} + \frac{\partial D_{mn}}{\partial \phi}\frac{d\tilde{\phi}}{d\alpha}\right]_{\alpha=0} + \cdots . \qquad (2.46)$$

To determine the infinitesimal operators A_r we have to determine the functions

$$\left[\frac{d\tilde{\psi}}{d\alpha}\right]_{\alpha=0}, \quad \left[\frac{d\tilde{\theta}}{d\alpha}\right]_{\alpha=0} \quad \text{and} \quad \left[\frac{d\tilde{\phi}}{d\alpha}\right]_{\alpha=0}, \qquad (2.47)$$

for each rotation.

Now the 3×3 matrix of the rotation g is a function of the angles ψ, θ and ϕ which, by Eq. (2.3), has the form

$$\begin{pmatrix}
\cos\psi & \sin^2\theta\cos\phi\sin\phi & \sin\theta\cos\theta\cos\phi \\
+\sin^2\theta\cos^2\phi & \times(1-\cos\psi) & \times(1-\cos\psi) \\
\times(1-\cos\psi) & -\cos\theta\sin\psi & +\sin\theta\sin\phi\sin\psi \\
\sin^2\theta\sin\phi\cos\phi & \cos\psi & \sin\theta\cos\theta\sin\phi \\
\times(1-\cos\psi) & +\sin^2\theta\sin^2\phi & \times(1-\cos\psi) \\
+\cos\theta\sin\psi & \times(1-\cos\psi) & -\sin\theta\cos\phi\sin\psi \\
\sin\theta\cos\theta\cos\phi & \sin\theta\cos\theta\sin\phi & \cos\psi \\
\times(1-\cos\psi) & \times(1-\cos\psi) & +\cos^2\theta(1-\cos\psi) \\
-\sin\theta\sin\phi\sin\psi & +\sin\theta\cos\phi\sin\psi &
\end{pmatrix}. \qquad (2.48)$$

The matrix of rotation gg' is given by some angles $\tilde{\psi}$, $\tilde{\theta}$ and $\tilde{\phi}$ which depend on the rotation angle α and which are equal to ψ, θ and ϕ when $\alpha = 0$. It will also be noted that expansion of the matrix gg' in a power series in α gives

$$gg' = g(\psi, \theta, \phi) + \alpha\left[\frac{\partial g}{\partial \psi}\left[\frac{d\tilde{\psi}}{d\alpha}\right]_{\alpha=0} + \frac{\partial g}{\partial \theta}\left[\frac{d\tilde{\theta}}{d\alpha}\right]_{\alpha=0} + \frac{\partial g}{\partial \phi}\left[\frac{d\tilde{\phi}}{d\alpha}\right]_{\alpha=0}\right] + \cdots . \qquad (2.49)$$

2.4. DIFFERENTIAL OPERATORS OF ROTATIONS

To find the infinitesimal operator A_1 we identify g' with the rotation with angle α around Ox_1 given by

$$g_1(\alpha) = \begin{pmatrix} 1 & 0 & 0 \\ 0 & \cos\alpha & -\sin\alpha \\ 0 & \sin\alpha & \cos\alpha \end{pmatrix}. \tag{2.50}$$

Therefore

$$g_1(\alpha) = \begin{pmatrix} 1 & 0 & 0 \\ 0 & 1 & 0 \\ 0 & 0 & 1 \end{pmatrix} + \alpha \begin{pmatrix} 0 & 0 & 0 \\ 0 & 0 & -1 \\ 0 & 1 & 0 \end{pmatrix} + \cdots. \tag{2.51}$$

As a consequence we obtain for the product of g with g_1:

$$gg_1 = g(\psi,\,\theta,\,\phi) + \alpha \begin{pmatrix} 0 & g_{13} & -g_{12} \\ 0 & g_{23} & -g_{22} \\ 0 & g_{33} & -g_{32} \end{pmatrix} + \cdots. \tag{2.52}$$

On the other hand, gg_1 is given by Eq. (2.49) when $g_1 = g'$. Comparing these two expressions for gg_1, we obtain equations from which the three expressions given in (2.47) can be determined for the case of rotation about Ox_1. We obtain

$$2\sin\theta\cos\phi\sin\frac{\psi}{2}\left(-\sin\theta\sin\phi\left[\frac{d\tilde\phi}{d\alpha}\right]_{\alpha=0} + \cos\theta\cos\phi\left[\frac{d\tilde\theta}{d\alpha}\right]_{\alpha=0}\right)$$

$$-\cos\frac{\psi}{2}\left(1 - \sin^2\theta\cos^2\phi\right)\left[\frac{d\tilde\psi}{d\alpha}\right]_{\alpha=0} = 0, \tag{2.53a}$$

$$2\sin\theta\sin\phi\sin\frac{\psi}{2}\left(\sin\theta\cos\phi\left[\frac{d\tilde\phi}{d\alpha}\right]_{\alpha=0} + \cos\theta\sin\phi\left[\frac{d\tilde\theta}{d\alpha}\right]_{\alpha=0}\right)$$

$$-\cos\frac{\psi}{2}\left(1 - \sin^2\theta\sin^2\phi\right)\left[\frac{d\tilde\psi}{d\alpha}\right]_{\alpha=0}$$

$$= \sin\theta\left(\cos\theta\sin\phi\sin\frac{\psi}{2} - \cos\phi\cos\frac{\psi}{2}\right), \tag{2.53b}$$

$$2\cos\theta\sin\frac{\psi}{2}\left[\frac{d\tilde\theta}{d\alpha}\right]_{\alpha=0} + \cos\frac{\psi}{2}\sin\theta\left[\frac{d\tilde\psi}{d\alpha}\right]_{\alpha=0}$$

$$= \cos\theta\sin\phi\sin\frac{\psi}{2} + \cos\phi\cos\frac{\psi}{2}. \tag{2.53c}$$

(One actually obtains nine equations; only three of them are independent. Equations (2.53) are obtained by equating the diagonal element of the matrices (2.49) and (2.52).)

2.4.3 Angular Momentum Operators

The solution of Eqs. (2.53) can easily be shown to be

$$\left[\frac{d\tilde\psi}{d\alpha}\right]_{\alpha=0} = \cos\phi\sin\theta, \tag{2.54a}$$

$$\left[\frac{d\tilde\theta}{d\alpha}\right]_{\alpha=0} = \frac{1}{2}\left(\sin\phi + \cos\frac{\psi}{2}\cos\theta\cos\phi\right), \tag{2.54b}$$

$$\left[\frac{d\tilde\phi}{d\alpha}\right]_{\alpha=0} = \frac{1}{2}\mathrm{cosec}\theta\left(\cos\theta\cos\phi - \cot\frac{\psi}{2}\sin\phi\right). \tag{2.54c}$$

Using Eqs. (2.54) in Eq. (2.46), we find the operator A_1 corresponding to the rotation around Ox_1:

$$A_1 = \cos\phi\sin\theta\frac{\partial}{\partial\psi} + \frac{1}{2}\left(\sin\phi + \cot\frac{\psi}{2}\cos\theta\cos\phi\right)\frac{\partial}{\partial\theta}$$

$$+ \frac{1}{2}\mathrm{cosec}\theta\left(\cos\theta\cos\phi - \cot\frac{\psi}{2}\sin\phi\right)\frac{\partial}{\partial\phi}. \tag{2.55a}$$

The operators A_2 and A_3 are found in a similar way:

$$A_2 = \sin\phi\sin\theta\frac{\partial}{\partial\psi} - \frac{1}{2}\left(\cos\phi - \cot\frac{\psi}{2}\cos\theta\sin\phi\right)\frac{\partial}{\partial\theta}$$

$$+ \frac{1}{2}\mathrm{cosec}\theta\left(\cos\theta\sin\phi + \cot\frac{\psi}{2}\cos\phi\right)\frac{\partial}{\partial\phi}, \tag{2.55b}$$

$$A_3 = \cos\theta\frac{\partial}{\partial\psi} - \frac{1}{2}\cot\frac{\psi}{2}\sin\theta\frac{\partial}{\partial\theta} - \frac{\partial}{\partial\phi}. \tag{2.55c}$$

Using the last three equations in Eq. (2.12) one obtains for the operators L_+, L_- and L_3:

$$L_\pm = ie^{\pm i\phi}\left[\sin\theta\frac{\partial}{\partial\psi} + \frac{1}{2}\left(\mp i + \cot\frac{\psi}{2}\cos\theta\right)\frac{\partial}{\partial\theta}\right]$$

$$+ie^{\pm i\phi}\left[\frac{1}{2}\operatorname{cosec}\theta\left(\cos\theta \pm i\cot\frac{\psi}{2}\right)\frac{\partial}{\partial\phi}\right], \qquad (2.56a)$$

$$L_3 = i\left(\cos\theta\frac{\partial}{\partial\psi} - \frac{1}{2}\cot\frac{\psi}{2}\sin\theta\frac{\partial}{\partial\theta} - \frac{\partial}{\partial\phi}\right). \qquad (2.56b)$$

The operators derived above were expressed in terms of the angle of rotation ψ and the spherical angles of direction of rotation θ and ϕ. One can use, however, the Euler angles and obtain the standard expressions of angular momentum operators given in the books of Naimark and Wigner.

In the following the Lorentz group is introduced along with its infinitesimal matrices and basic infinitesimal operators. The commutation relations that these matrices and operators satisfy are also given. This is the infinitesimal approach to finding the representation of the Lorentz group. Each representation is shown to be completely determined by a pair of numbers.

2.5 Infinite-Dimensional Representations

In this section a brief review of the theory of infinite-dimensional representations is given. For more details the reader is referred to the books of Naimark and of Gelfand *et al.*

2.5.1 Banach Space

A linear space is called *Euclidean* if in it a function (x, y), called the *scalar product* of x and y, is defined and satisfies the following:
(1) $(x, x) \geq 0$, $(x, x) = 0$ if and only if $x = 0$;
(2) $(y, x) = \overline{(x, y)}$;
(3) $(\alpha x, y) = \alpha (x, y)$;
(4) $(x_1 + x_2, y) = (x_1, y) + (x_2, y)$,
for any number α.

Normed Space

A linear space R is said to be *normed* if a function, denoted by $\mid x \mid$, is defined in it, which satisfies:
(1) $\mid x \mid \geq 0$, $\mid x \mid = 0$ if and only if $x = 0$;
(2) $\mid \alpha x \mid = \mid \alpha \mid \mid x \mid$ for any number α and any $x \in R$;
(3) $\mid x + y \mid \leq \mid x \mid + \mid y \mid$ for any $x, y \in R$.
Such a function $\mid x \mid$ is called a *norm*.

An example of a normed space is the aggregate C of all complex numbers x. The norm of a complex number is taken as its modulus. Another example is provided by the aggregate of all sequences $x = \{\xi_1, \xi_2, ...\}$ of complex numbers $\xi_1, \xi_2, ...$ for which the series $|\xi_1|^2 + |\xi_2|^2 + \cdots$ converges. The operations in the space are defined as $\alpha x = \{\alpha\xi_1, \alpha\xi_2, ...\}$ and $x + y = \{\xi_1 + \eta_1, \xi_2 + \eta_2, ...\}$ for $x = \{\xi_1, \xi_2, ...\}$ and $y = \{\eta_1, \eta_2, ...\}$. The norm is defined as $|x| = \{|\xi_1|^2 + |\xi_2|^2 + \cdots\}^{1/2}$. This space is sometimes denoted by l^2.

Let R be a Euclidean space, not necessarily finite-dimensional. Then, in R, a norm can be defined by

$$|x| = \sqrt{(x,x)}. \tag{2.57}$$

The axioms for a norm will be satisfied; in fact, the first two axioms are trivially satisfied. To prove the triangle inequality one needs

$$|(x,y)| \leq |x| \, |y|, \tag{2.58}$$

which is the well-known Cauchy-Buniakovsky inequality. From the above inequality one has

$$|x+y|^2 = (x+y, x+y) = (x,x) + (x,y) + \overline{(x,y)} + (y,y)$$
$$\leq |x|^2 + 2|x|\,|y| + |y|^2 = (|x| + |y|)^2, \tag{2.59}$$

and consequently

$$|x+y| \leq |x| + |y|. \tag{2.60}$$

A sequence of elements x_n of a normed space R is called *convergent in norm* to the element x of R if $|x - x_n| \to 0$ as $n \to \infty$. A sequence x_n of R is called *fundamental* if it satisfies the Cauchy condition (i.e. if for every $\epsilon > 0$ there exists a number $N = N(\epsilon)$ such that $|x_n - x_m| < \epsilon$ for $n, m > N$.) A space R is called *complete* if every fundamental sequence in R converges in norm to some element x of R. A complete normed space is called a *Banach space*.

Examples of complete normed spaces are the space C of all complex numbers and the space l^2, both mentioned above. An example of a non-complete normed space is the set of all sequences $x = \{\xi_1, \xi_2, \cdots\}$ in which only a finite number of ξ_n is non-zero, all other operations of the space are the same as those of the space l^2.

Let S be an arbitrary set in a Banach space R. The set $\overline{S}$ obtained from S by adding to it all the limits in norm of sequences of elements x_n of S is

2.5. INFINITE-DIMENSIONAL REPRESENTATIONS

called the *closure* of the set S. A set S is called *dense* in R if $\overline{S} = R$. A set S is called *closed* if $\overline{S} = S$. A closed subspace of a Banach space is itself a Banach space.

A series $x_1 + x_2 + \cdots$ of elements x_n of R is called *convergent* and the element x of R is called the sum of the series if $x_1 + x_2 + \cdots + x_n \to x$ as $n \to \infty$ in the sense of the norm in R. A series $x_1 + x_2 + \cdots$ is called *absolutely convergent* if the series $\mid x_1 \mid + \mid x_2 \mid + \cdots$ of real numbers is convergent. In a Banach space every absolutely convergent series converges. This follows from the inequality

$$\mid x_{n+1} + \cdots + x_{n+p} \mid \leq \mid x_{n+1} \mid + \cdots + \mid x_{n+p} \mid \tag{2.61}$$

and the fact that the space is complete.

2.5.2 Hilbert Space

A Euclidean space R, complete with respect to the norm $\mid x \mid = \sqrt{(x,x)}$, is called a *Hilbert space*.

Examples of Hilbert Spaces

The space l^2 discussed in Subsection 2.5.1 is a Hilbert space if the scalar product is defined by $(x,y) = \sum \xi_k \overline{\eta}_k$ for $x = \{\xi_1, \xi_2, \cdots\}$ and $y = \{\eta_1, \eta_2, \cdots\}$. Another example of a Hilbert space is the aggregate of all functions $f(x)$, measurable in a fixed interval (a,b) and satisfying the conditions $\int_a^b \mid f(x) \mid^2 dx < \infty$, if the operations of addition and multiplication by a number are defined in the usual way, and the scalar product is defined by

$$(f_1, f_2) = \int_a^b f_1(x) \overline{f_2}(x) \, dx.$$

This Hilbert space is sometimes denoted by $L^2(a,b)$. In the same way the Hilbert space $L^2(SU(2))$ is defined as the aggregate of all functions $f(u)$ satisfying $\int \mid f(u) \mid^2 du < \infty$, where the scalar product is defined by

$$(f_1, f_2) = \int f_1(u) \overline{f_2}(u) \, du.$$

A *linear functional* $f(x)$ in a linear space R is a numerical function satisfying $f(\alpha x) = \alpha f(x)$ and $f(x+y) = f(x) + f(y)$ for any number α and x and y of R. A linear functional in a normed space R is called *bounded* if there exists a constant $c \geq 0$ such that $\mid f(x) \mid \leq c \mid x \mid$ for all

x of R. The smallest number $c \geq 0$ satisfying this condition is called the norm of the functional and is denoted by $\mid f \mid$. Thus $\mid f(x) \mid \leq \mid f \mid \mid x \mid$. The bounded linear functionals in R form a normed linear space where the sum and product are defined by $(f_1 + f_2)(x) = f_1(x) + f_2(x)$ and $(\alpha f)(x) = \overline{\alpha} f(x)$. This space is called the conjugate to the space R. It is a complete space.

In a Hilbert space R every bounded linear functional $f(x)$ is represented in the form $f(x) = (x, y)$, where y belongs to R and $\mid f \mid = \mid y \mid$. Hence the space R', conjugate to a Hilbert space R, may be identified with R itself, $R' = R$.

Two Hilbert spaces R_1 and R_2 are called *isometric* if there exists a linear operator U mapping R_1 onto R_2 and preserving the scalar product, $(Ux, Uy) = (x, y)$ for all x, y of R_1. The operator U itself is then called *isometric* and it satisfies $\mid Ux \mid = \mid x \mid$ for all x of R_1.

2.5.3 Operators in a Banach Space

A linear operator A in a Banach space R is called bounded if there exists a constant $c \geq 0$ such that $\mid Ax \mid \leq c \mid x \mid$ for all x of R. The smallest number c satisfying this condition is called the *norm* of the bounded operator A and is denoted by $\mid A \mid$. Hence $\mid Ax \mid \leq \mid A \mid \mid x \mid$. If A and B are bounded operators, then also the operators αA, $A + B$, and BA are bounded and satisfy

$$\mid \alpha A \mid = \mid \alpha \mid \mid A \mid; \quad \mid A + B \mid \leq \mid A \mid + \mid B \mid; \quad \mid AB \mid \leq \mid A \mid \mid B \mid . \quad (2.62)$$

It then follows that every bounded linear operator A is continuous. Furthermore, if two bounded operators A and B coincide on a set S which is dense in a space R, then they coincide on the whole of R.

2.5.4 General Definition of a Representation

A mapping $g \to D(g)$ of a group G on a Banach space R is called a *representation* if to every element g of G there corresponds a bounded linear operator $D(g)$ in R such that $D(e) = 1$ and $D(g_1 g_2) = D(g_1) D(g_2)$. A representation $g \to D(g)$ in a Banach space R is called *irreducible* if R contains no closed subspace (other than the null one and R itself) which is invariant with respect to all operators $D(g)$. This definition coincides with that of irreducibility for the finite-dimensional one. This is so since every finite-dimensional subspace is closed.

2.5. INFINITE-DIMENSIONAL REPRESENTATIONS

Let a linear space be denoted by R and let its conjugate space be denoted by R'. Then for every element x of R there exists a functional f of R' such that $f(x) = |x|$ and $|f| = 1$. Hence if $f(x) = 0$ for all f of R' then $x = 0$. Furthermore, if M is a closed subspace of a Banach space R and x_0 is a vector in R not belonging to M, then there exists a functional f of R' satisfying $f(x_0) \neq 0$, and $f(x) = 0$ for all x of M.

Since the conjugate space R' is a normed space, one can therefore consider the linear bounded functional $F(f)$ in it. Such functionals are obtained, for example, if we put $F_x(f) = \overline{f(x)}$ for a fixed element x of R since $F_x(f)$ is a bounded linear functional in R'. A space R is called reflexive if the functionals $F_x(f)$, for all x of R, exhaust all the bounded linear functionals in R'. In other words if every bounded linear functional $F(f)$ in R' is given by $F(f) = f(x)$ for some x of R. (Throughout our discussion we consider only representations in reflexive Banach spaces.)

2.5.5 Continuous Representations

Let $x(t) = x(t_1, t_2, \cdots, t_m)$ be a vector function of a point $t = (t_1, \cdots, t_m)$ in an m-dimensional space with values in R. A vector function $x(t)$ is called *continuous* in a set D in m-dimensional space if for every functional f of the conjugate space R' the numerical function $f[x(t)]$ is continuous in D. A bounded linear operator function $A(t)$ in R is called *continuous* in D if for every x of R and f of R' the numerical function $f(a(t)x)$ is continuous in D. For example if G is a group of matrices, then it may be regarded as a subset of m-dimensional space for a sufficiently large m. Hence one may speak of a vector-function $x(g)$ or an operator function $A(g)$ as continuous in the group G.

A representation $g \to D(g)$ of a group of matrices is called continuous if $D(g)$ is a continuous operator function. [Throughout the text the term representation stands for continuous representation (unless otherwise stated)]. It then follows that if $x(t)$ and $A(t)$ are vector and operator functions, respectively, which are continuous in a closed bounded set D, then the numerical functions $|x(t)|$ and $|A(t)|$ are bounded in that set.

2.5.6 Unitary Representations

The concept of a unitary representation in finite-dimensional spaces discussed in Subsection 2.1.3 can be generalized to infinite dimensions as follows.

Let U be an isometric operator mapping R onto itself, then U is called a *unitary operator* in R. A representation $g \to D(g)$ of a group G in a space R is called *unitary* if R is a Hilbert space and $D(g)$ is a unitary operator for all g of G. A representation $g \to D(g)$ in a Hilbert space R is unitary if

$$(D(g)x, D(g)y) = (x,y) \tag{2.63}$$

for all g of G and x, y of R.

Now let A be a bounded operator in a Hilbert space R. An operator $A^\dagger$ is called *adjoint* to A if $(Ax,y) = (x, A^\dagger y)$ for all x, y of R. One can show that $A^{\dagger\dagger} = A$, $(\alpha A)^\dagger = \alpha A^\dagger$, $(A+B)^\dagger = A^\dagger + B^\dagger$, $(AB)^\dagger = B^\dagger A^\dagger$, and $|A^\dagger| = |A|$. An operator U is unitary if and only if $U^\dagger U = UU^\dagger = 1$. The operator A^{-1} is called *inverse* to A if $AA^{-1} = A^{-1}A = 1$. Hence a unitary operator satisfies $U^\dagger = U^{-1}$. An operator A is called Hermitian if $A^\dagger = A$. A Hermitian operator P is called a *projection* operator if $P^2 = P$.

Finally, let $R_1, R_2, \cdots$ be closed, mutually orthogonal, subspaces of a Hilbert space R. The aggregate of all sums $x = x_1 + x_2 + \cdots$ of convergent series of elements $x_k \in R_k$ is called the *orthogonal sum* of the Hilbert spaces $R_1, R_2, \cdots$, and is denoted by $R_1 \bigoplus R_2 \bigoplus \cdots$. It follows that $R_1 \bigoplus R_2 \bigoplus \cdots$ is a closed subspace of R. If E_n is a projection operator in R onto $R_1 \bigoplus R_2 \bigoplus \cdots \bigoplus R_n$, then $E_n x = x_1 + \cdots + x_n$ for any vector $x = x_1 + x_2 + \cdots$ of R, where $x_k \in R_k$.

The bounded linear operator A in a space R is called the *orthogonal sum* of the operators A_k in R_k, denoted by $A_1 \bigoplus A_2 \bigoplus \cdots$, if $R = R_1 \bigoplus R_2 \bigoplus \cdots$ and $Ax = A_1 x_1 + A_2 x_2 + \cdots$, where $x = x_1 + x_2 + \cdots$. A unitary representation $g \to D(g)$ of a group G in a Hilbert space R is called the orthogonal sum of the representations $g \to D^{(k)}(g)$ in the closed subspaces R_k if $D(g) = D^{(1)}(g) + D^{(2)}(g) + \cdots$ for all g of G.

In the next chapter the Lorentz group and the group SL(2,C) are discussed in details.

2.6 References for Further Reading

M. Carmeli, Representations of the three-dimensional rotation group in terms of direction and angle of rotation, *J. Math. Phys.* **9**, 1987-1992 (1968). (Sections 2.2, 2.4)

M. Carmeli and S. Malin, *Representations of the Rotation and Lorentz Groups* (Marcel Dekker, New York and Basel, 1976).

I.M. Gelfand, M.I. Graev and I.I. Pyatetskii-Shapiro, *Representation Theory and Automorphic Functions* (W.B. Saunders Co., Philadelphia, (1969).

2.6. REFERENCES FOR FURTHER READING

(Section 2.5)

I.M. Gelfand, M.I. Graev and N.Ya. Vilenkin, *Generalized Functions, Vol.5: Integral Geometry and Representation Theory* (Academic Press, New York, 1966) (Sections 2.1, 2.5).

I.M. Gelfand, R.A. Minlos and Z.Ya. Shapiro, *Representations of the Rotation and Lorentz Groups and their Applications* (Pergamon Press, New York, 1963). (Section 2.1)

I.M. Gelfand and Z.Ya. Shapiro, Representations of the group of rotations in three-dimensional space and their applications, *Usp. Mat. Nauk* **7**, 3-117 (1952); *Amer. Math. Soc. Transl. Ser.2.*, **2**, 207-316 (1956). (Section 2.1)

H.E. Moses, Irreducible representations of the rotation group in terms of the axis and angle of rotation, *Ann. Phys.(N.Y)* **37**, 224-226 (1966); **42**, 343-346 (1967). (Section 2.2)

H.E. Moses, Irreducible representations of the rotation group in terms of Euler's theorem, *Nuovo Cimento* **40A**, 1120-1138 (1965). (Section 2.2)

M.A. Naimark, *Linear Representations of the Lorentz Group* (Pergamon Press, New York, 1964). (Sections 2.1, 2.4, 2.5)

L.S. Pontrjagin, *Topological Groups* (Princeton University Press, Princeton, New Jersey, 1946). (Section 2.3)

E.P. Wigner, *Group Theory and its Application to the Quantum Mechanics of Atomic Spectra* (Academic Press, New York, 1964). (Section 2.4)

Chapter 3

The Lorentz and SL(2,C) Groups

In this chapter the elements of the theory of special relativity are given and the relationship between the Lorentz group and the group SL(2,C) is discussed extensively. We first discuss the Lorentz group. The group SL(2,C), the aggregate of all 2×2 complex matrices with determinant unity, is consequently introduced. It is shown that SL(2,C) is homomorphic to the homogeneous, orthochronous, Lorentz group. A direct correspondence between elements of the matrices of the two groups is given explicitly. The representations of these groups are given in the next chapter.

3.1 Elements of Special Relativity

In this section the fundamentals of Einstein's special relativity theory are given. It is within this theory that the Lorentz group originates.

3.1.1 Postulates of Special Relativity

In the following we give the basic principles of the special theory of relativity. These principles are needed to describe the electromagnetic field and other physical phenomena, and they constitute their spacetime symmetry background.

The special theory of relativity was developed by Einstein in 1905 in order to overcome and correct certain basic concepts that were in use at that time, such as asymmetries in relative motion of bodies. Examples of relative motion in electrodynamics, and the unsuccessful attempt to detect the motion of the Earth by the experiment of Michelson and Morley, suggested that the phenomena of electrodynamics and mechanics do not depend on the Newtonian notion of absolute rest. Rather, the laws of electrodynamics should be valid in all frames of references in which the equations of mechanics are valid.

Einstein raised the above observation to the status of a postulate and called it *the principle of relativity*. He also introduced another postulate (which is only apparently inconsistent with the former one) according to which light always propagates in empty space with a constant velocity c which is independent of the motion of the emitting body and the measuring instrument.

The above two postulates were shown by Einstein to be enough for the development of a consistent theory of electrodynamics of moving charges which is based on Maxwell's original theory that was assumed to be valid in stationary systems only. The theory did not require an "absolute stationary space."

To describe the electromagnetic field, or any other classical field, one needs a system of coordinates in terms of which the fields are described. Such a coordinate system will include three *spatial* coordinates to which we add the *time* coordinate. The three spatial coordinates will be denoted by x^k, where lower case Latin indices $k = 1, 2, 3$, and the time coordinate by $x^0 = ct$, where c is the speed of light in vacuum. The four coordinates will collectively be denoted by x^α, where Greek indices take the values $\alpha = 0, 1, 2, 3$.

Inertial Coordinate System

A system of coordinates in which the law of inertia holds is called an *inertial coordinate system*. Hence Newton's laws of mechanics are valid only in inertial coordinate systems.

If K is an inertial coordinate system, then every other coordinate system K' is also an inertial system if it is in uniform motion with respect to K. Hence if, relative to K, K' is a uniformly moving coordinate system then the physical laws can be expressed with respect to K' exactly as with respect to K.

One of the most important physical consequences of the special relativity theory is the existence of a maximum signal speed in nature, which coincides with the velocity of light in empty space. It is therefore natural to define

3.1. ELEMENTS OF SPECIAL RELATIVITY

the same time at separate points by means of light signals. This then raises the problem of defining simultaneity.

Simultaneity

The definition of simultaneity is made as follows. If light requires the same time to pass across a path $A \to M$ as for a path $B \to M$, where M is the middle of the distance AB, then we say that the light signals at A and B started simultaneously if the observer at M sees the two light signals at the same time.

Will the two events, which occur simultaneously in one system, also be simultaneous in another system moving with a velocity v with respect to the first one? The answer is *negative*; events which are simultaneous in one coordinate system are not necessarily simultaneous in others. It follows that every inertial system has its own particular time.

3.1.2 The Galilean Transformation

As was mentioned above, inertial coordinate systems are those which are in uniform, rectilinear, translational motions with respect to each other. Accordingly, inertial systems of coordinates differ from each other by orthogonal rotations, accompanied by translations of the origins of the systems, and by motion with uniform velocities. One can, furthermore, add the translation of the time coordinate thus enabling an arbitrary choice of the origin of time $t = 0$.

Counting the number of parameters which each system of coordinates has with respect to any other, we find that there are ten.

A transformation between inertial coordinate systems which has ten parameters, as described above, is called a *Galilean transformation*. The aggregate of all Galilean transformations provides a group, called the *Galilean group*, which has ten parameters.

One can choose two inertial systems of coordinates so that their corresponding axes are parallel and coincide at $t = 0$. If v is the velocity of one inertial coordinate system with respect to the other, the Galilean transformation can then be reduced into a simple transformation as follows:

$$x' = x - v_x t, \tag{3.1a}$$

$$y' = y - v_y t, \tag{3.1b}$$

$$z' = z - v_z t, \tag{3.1c}$$

where v_x, v_y and v_z are the components of the velocity **v** along the x axis, y axis, and z axis, respectively. Of course the Newtonian laws of classical mechanics are invariant under the full ten-parameter Galilean group of trasnformations, and we have what can be called a *Galilean invariance*.

3.1.3 The Lorentz Transformation

According to the Galilean transformation which relates the spatial coordinates and the time between inertial systems in prerelativity, the postulates of the constancy of the speed of light and of the principle of relativity (which applies, in particular, to the propagation of light and hence its constant velocity is independent of the choice of the inertial system) are mutually incompatible, even though both are experimentally valid.

The special theory of relativity resolves this impasse as follows.

The above two postulates will be compatible with each other if a new transformation relating the spatial coordinates and times of different inertial systems replaces the Galilean transformation. The new transformation, of course, follows to be the Lorentz transformation. This, subsequently, requires certain behavior of the moving measuring rods and clocks.

The principle of relativity may, thus, alternatively be restated as follows: *The laws of physics should be covariant (or invariant) under the Lorentz transformations relating different inertial coordinate systems.* This Lorentz invariance is in accordance with the Michelson-Morley null experiment which showed that on the moving Earth light spreads with the same speeds in all directions.

Consequently, the behavior of light is not incompatible with the principle of relativity. The incompatibility is only apparent.

3.1.4 Derivation of the Lorentz Transformation

We now derive the Lorentz transformation connecting the two coordinate systems K and K' when they have the same orientations and their origins coincide at $t = 0$, but K' moves along the coordinate x with a velocity v.

The directions perpendicular to the motion are obviously left unaffected by the transformation. Hence

$$x'^2 = x^2, \ x'^3 = x^3, \tag{3.2}$$

3.1. ELEMENTS OF SPECIAL RELATIVITY

and only the x^0 and x^1 coordinates require changes when transforming from one system to the other. One will therefore have the form

$$\Lambda = \begin{pmatrix} \Lambda^0{}_0 & \Lambda^0{}_1 & 0 & 0 \\ \Lambda^1{}_0 & \Lambda^1{}_1 & 0 & 0 \\ 0 & 0 & 1 & 0 \\ 0 & 0 & 0 & 1 \end{pmatrix} \quad (3.3)$$

for the matrix of the Lorentz transformation in our particular case.

The "orthogonality" condition of the Lorentz transformation yields

$$\eta_{CD}\Lambda^C{}_A\Lambda^D{}_B = \eta_{AB}, \quad (3.4)$$

where the indices A, B, C, $D = 0$, 1, and $\eta_{00} = -\eta_{11} = 1$, $\eta_{01} = \eta_{10} = 0$. The above formula gives three relations connecting the four elements of the matrix (3.3):

$$\left(\Lambda^0{}_0\right)^2 - \left(\Lambda^1{}_0\right)^2 = 1,$$

$$\left(\Lambda^0{}_1\right)^2 - \left(\Lambda^1{}_1\right)^2 = -1, \quad (3.5)$$

$$\Lambda^0{}_0\Lambda^0{}_1 - \Lambda^1{}_0\Lambda^1{}_1 = 0.$$

The solution of these equations can therefore be determined up to an arbitrary parameter. One then finds that

$$\begin{aligned}\Lambda^0{}_0 &= \cosh\psi, \quad \Lambda^0{}_1 = \sinh\psi, \\ \Lambda^1{}_0 &= \sinh\psi, \quad \Lambda^1{}_1 = \cosh\psi.\end{aligned} \quad (3.6)$$

is such an appropriate solution. With these values for the four elements, we obtain

$$\Lambda = \begin{pmatrix} \cosh\psi & \sinh\psi & 0 & 0 \\ \sinh\psi & \cosh\psi & 0 & 0 \\ 0 & 0 & 1 & 0 \\ 0 & 0 & 0 & 1 \end{pmatrix} \quad (3.7)$$

for the matrix (3.3) of the Lorentz transformation.

The parameter ψ is related to the relative velocity v between the two inertial coordinate systems K and K'. The relationship between them is found by determining the motion of the origin of the coordinate system K as seen from K', for instance. This motion is determined by putting $x^1 = 0$ in the Lorentz transformation, and using Eq. (3.7). This gives

$$\begin{aligned}x'^0 &= x^0\cosh\psi, \\ x'^1 &= x^0\sinh\psi.\end{aligned} \quad (3.8)$$

We therefore obtain

$$\frac{x'^1}{x'^0} = \frac{1}{c}\frac{x'}{t'} = -\beta = \tanh\psi, \qquad (3.9)$$

where the parameter β is defined by

$$\beta = \frac{v}{c}. \qquad (3.10)$$

Accordingly we obtain from Eq. (3.9)

$$\cosh\psi = \frac{1}{\sqrt{1-\beta^2}}, \qquad (3.11a)$$

$$\sinh\psi = \frac{-\beta}{\sqrt{1-\beta^2}}. \qquad (3.11b)$$

Using these results in Eq. (3.7) then yields

$$\Lambda = \begin{pmatrix} \frac{1}{\sqrt{1-\beta^2}} & \frac{-\beta}{\sqrt{1-\beta^2}} & 0 & 0 \\ \frac{-\beta}{\sqrt{1-\beta^2}} & \frac{1}{\sqrt{1-\beta^2}} & 0 & 0 \\ 0 & 0 & 1 & 0 \\ 0 & 0 & 0 & 1 \end{pmatrix} \qquad (3.12)$$

for the matrix of the Lorentz transformation. We also obtain

$$\Lambda^{-1} = \begin{pmatrix} \frac{1}{\sqrt{1-\beta^2}} & \frac{\beta}{\sqrt{1-\beta^2}} & 0 & 0 \\ \frac{\beta}{\sqrt{1-\beta^2}} & \frac{1}{\sqrt{1-\beta^2}} & 0 & 0 \\ 0 & 0 & 1 & 0 \\ 0 & 0 & 0 & 1 \end{pmatrix} \qquad (3.13)$$

for the inverse matrix describing the inverse Lorentz transformation.

The Lorentz transformation along the x axis is therefore given by

$$ct' = \frac{ct - \beta x}{\sqrt{1-\beta^2}}, \qquad (3.14a)$$

$$x' = \frac{x - \beta ct}{\sqrt{1-\beta^2}}, \qquad (3.14b)$$

3.1. ELEMENTS OF SPECIAL RELATIVITY

$$y' = y, \quad z' = z. \tag{3.14c}$$

We also obtain

$$ct = \frac{ct' + \beta x'}{\sqrt{1 - \beta^2}}, \tag{3.15a}$$

$$x = \frac{x' + \beta ct'}{\sqrt{1 - \beta^2}}, \tag{3.15b}$$

$$y = y', \quad z = z'. \tag{3.15c}$$

for the inverse transformation from the coordinates x'^{μ} back to x^{μ}.

Equations (3.15) show that the inverse transformation differs from Eqs. (3.14) only by a change in the sign of v. This result is obvious since the coordinate system K is moving relative to the system K' with the velocity $-v$.

A Lorentz transformation involving the time coordinate x^0 and one or more spatial coordinates x^k, such as that derived above, is often called a boost. A Lorentz transformation which keeps the time coordinate unchanged is, of course, just an ordinary three-dimensional rotation of the spatial coordinates.

3.1.5 The Cosmological Transformation

Universe Expansion versus Light Propagation

The Lorentz transformation has an analog in cosmology. (For full details see Carmeli's book appearing in the References at the end of this chapter.) Under the assumption that gravitation is negligible and thus Hubble's constant is constant in cosmic time, there is an analogy between the propagation of light, $x = ct$, and the expansion of the Universe, $x = \tau v$, where τ is Hubble's time, a constant which is also the age of the Universe under the above assumption, and c is the speed of light in vacuum.

Thus one can express the expansion of the Universe, assuming that it is homogeneous and isotropic, in terms of the null vector (v, x, y, z) satisfying

$$\tau^2 v^2 - \left(x^2 + y^2 + z^2\right) = 0, \tag{3.16}$$

where v is the receding velocity of the galaxies. Equation (3.16), in the 4-dimensional flat space of the Cartesian 3-space and the velocity, is similar to

$$c^2 t^2 - \left(x^2 + y^2 + z^2\right) = 0, \tag{3.17}$$

describing the null propagation of light in Minkowskian spacetime. We assume, furthermore, that a relationship of the form (3.16) is valid at all cosmic times, just as the assumption that (3.17) holds in all inertial systems moving at any speed.

Accordingly, at a cosmic time t' at which the coordinates and velocity are labeled with primes, we have

$$\tau^2 v'^2 - (x'^2 + y'^2 + z'^2) = 0, \qquad (3.18)$$

with the same τ, just as for light emitted from a source with velocity v with respect to the first one,

$$c^2 t'^2 - (x'^2 + y'^2 + z'^2) = 0. \qquad (3.19)$$

As a result, we have a 4-dimensional space with zero curvature of v, x, y, z just as the Minkowskian spacetime of t, x, y, z.

We now assume that at two cosmic times t and t' we have

$$\tau^2 v'^2 - (x'^2 + y'^2 + z'^2) = \tau^2 v^2 - (x^2 + y^2 + z^2), \qquad (3.20)$$

in analogy to the special relativistic formula

$$c^2 t'^2 - (x'^2 + y'^2 + z'^2) = c^2 t^2 - (x^2 + y^2 + z^2). \qquad (3.21)$$

The question is then what is the transformation between v', x', y', z' and v, x, y, z that satisfies the invariance formula (3.20).

The transformation can be derived like deriving the Lorentz transformation by writing, in the two-dimensional case,

$$\tau^2 v'^2 - x'^2 = \tau^2 v^2 - x^2, \qquad (3.22)$$

whose solution is

$$\begin{aligned} \tau v' &= \tau v \cosh \psi - x \sinh \psi, \\ x' &= x \cosh \psi - \tau v \sinh \psi. \end{aligned} \qquad (3.23)$$

At $x' = 0$ we obtain

$$\tanh \psi = \frac{x}{\tau v} = \frac{t}{\tau}, \qquad (3.24)$$

and therefore

$$\sinh \psi = \frac{t/\tau}{\sqrt{1 - \frac{t^2}{\tau^2}}}, \qquad (3.25a)$$

3.1. ELEMENTS OF SPECIAL RELATIVITY

$$\cosh \psi = \frac{1}{\sqrt{1 - \frac{t^2}{\tau^2}}}, \qquad (3.25b)$$

which lead to the transformation

$$v' = \frac{v - xt/\tau^2}{\sqrt{1 - \frac{t^2}{\tau^2}}}, \qquad (3.26a)$$

$$x' = \frac{x - tv}{\sqrt{1 - \frac{t^2}{\tau^2}}}, \qquad (3.26b)$$

$$y' = y, \quad z' = z. \qquad (3.26c)$$

The cosmic time appearing in the transformations (3.26) is measured backward with respect to us. Thus the present time is zero, and that at the big bang is τ.

The transformation (3.26) is called the *cosmological transformation* and it relates physical quantities at different cosmic times. It is in complete analogy to the Lorentz transformation which relates physical quantities at different velocities.

Interpretation of the Cosmological Transformation

Equations (3.26) give the transformed values of x and v as measured in the system K' with a relative cosmic time t with respect to K. The roles of the time and the velocity are *exchanged* as compared to special relativity. This fits our needs in cosmology where one measures distances and velocities at different cosmic times in the past. The parameter t/τ replaces v/c of special relativity.

It should be emphasized that the transformation (3.26) is not a trivial exchange of v/c, appearing in the Lorentz transformation, and t/τ here. For example, the redshift $z = v/c$ at low velocities, but is certainly not equal to t/τ for small t/τ.

The Galaxy Cone

The invariant equation (16), describing the distribution of galaxies in the Universe at any cosmic time, has a very simple geometrical interpretation. It enables one to present the locations of galaxies as a cone in the dual space of distance and velocity. One then has a *galaxy cone*, similar to the familiar light cone in special relativity. The symmetry axis of the cone coincides with the x^0 axis which extends from $-\tau c$ to $+\tau c$.

3.2 The Lorentz Group

Consider two inertial coordinate systems K and K' whose origins coincide at time $t = 0$, and the events with respect to which are denoted by t, x, y, z and t', x', y', z', respectively. A light pulse emitted from the origin of K will be spread spherically with the speed c, according to the equation

$$x^2 + y^2 + z^2 = c^2 t^2. \tag{3.27}$$

Invariance of the speed of light tells us that an observer in K' will also see the light propagating from his origin spherically according to the equation

$$x'^2 + y'^2 + z'^2 = c^2 t'^2. \tag{3.28}$$

From Eqs. (3.27) and (3.28) one then obtains

$$c^2 t'^2 - (x'^2 + y'^2 + z'^2) = c^2 t^2 - (x^2 + y^2 + z^2), \tag{3.29}$$

or

$$\eta_{\mu\nu} x'^\mu x'^\nu = \eta_{\mu\nu} x^\mu x^\nu, \tag{3.30}$$

where x^μ and x'^μ are defined by

$$x^\mu = (ct,\ x,\ y,\ z),\ x'^\mu = (ct',\ x',\ y',\ z'), \tag{3.31}$$

and the symbol $\eta_{\mu\nu}$ (and later on $\eta^{\mu\nu}$) is the flat-space metric, given by the matrix

$$\eta = \begin{pmatrix} +1 & 0 & 0 & 0 \\ 0 & -1 & 0 & 0 \\ 0 & 0 & -1 & 0 \\ 0 & 0 & 0 & -1 \end{pmatrix}. \tag{3.32}$$

In the above equations, and throughout the following, repeated indices indicate the use of the summation convention.

We will seek a linear transformation of the form

$$x'^\mu = \Lambda^\mu{}_\nu x^\nu \tag{3.33}$$

between the times and spatial coordinates of the two inertial systems K and K'. Using matrix notation, Eqs. (3.30) and (3.33) can then be written in the form

$$x'^t \eta x' = x^t \eta x \tag{3.34}$$

3.2. THE LORENTZ GROUP

and
$$x' = \Lambda x, \qquad (3.35)$$

respectively. Here x and x' are the one-column matrices

$$x = \begin{pmatrix} x^0 \\ x^1 \\ x^2 \\ x^3 \end{pmatrix}, \quad x' = \begin{pmatrix} x'^0 \\ x'^1 \\ x'^2 \\ x'^3 \end{pmatrix}, \qquad (3.36)$$

x^t and x'^t are the transposed matrices to the matrices x and x', respectively, and Λ is the 4×4 matrix whose elements are $\Lambda^\mu{}_\nu$.

Using now Eq. (3.35) in Eq. (3.34) then gives

$$x^t \Lambda^t \eta \Lambda x = x^t \eta x, \qquad (3.37)$$

from which we obtain the condition

$$\Lambda^t \eta \Lambda = \eta \qquad (3.38)$$

that the 4×4 matrix Λ of the transformation has to satisfy. The transformation (3.33) is the *Lorentz transformation*. Equation (3.38) is a generalization of the familiar relation

$$R^t I R = I, \qquad (3.39)$$

which the 3×3 orthogonal matrix R, describing ordinary rotations of the spatial coordinates alone, satisfies. The essential difference between the two cases is in the replacement of the unit matrix I in the ordinary three-dimensional rotations by the matrix η in the four-dimensional Lorentz transformations.

Equation (3.38) shows that

$$(\det \Lambda)^2 = 1, \qquad (3.40)$$

and accordingly the determinant of every Lorentz transformation Λ is equal to either $+1$,

$$\det \Lambda = +1, \qquad (3.41)$$

in which case the transformation is called *proper*, or to -1,

$$\det \Lambda = -1, \qquad (3.42)$$

where the transformation is called *improper*. This is similar to the case of the rotation group discussed in the previous chapter.

3.2.1 Orthochronous Lorentz Transformation

From Eq. (3.12), when written with indices, one finds

$$\eta_{\mu\nu}\Lambda^{\mu}{}_{\alpha}\Lambda^{\nu}{}_{\beta} = \eta_{\alpha\beta}, \qquad (3.43)$$

and taking $\alpha = \beta = 0$, one obtains

$$\left(\Lambda^{0}{}_{0}\right)^2 - \left(\Lambda^{1}{}_{0}\right)^2 - \left(\Lambda^{2}{}_{0}\right)^2 - \left(\Lambda^{3}{}_{0}\right)^2 = 1. \qquad (3.44)$$

Therefore $\left(\Lambda^{0}{}_{0}\right)^2 \geq 1$, and consequently we have either

$$\Lambda^{0}{}_{0} \geq +1, \qquad (3.45)$$

in which case the transformation is called *orthochronous*, or

$$\Lambda^{0}{}_{0} \leq -1. \qquad (3.46)$$

The aggregate of all orthochronous Lorentz transformations provides a subgroup of the Lorentz group. The aggregate of all proper, orthochronous, Lorentz transformations also provides a group which is a subgroup of the Lorentz group.

In the following we will be concerned with the group of all proper, orthochronous, Lorentz transformations. This group will be denoted by L.

3.2.2 Subgroups of the Lorentz Group

To conclude this section we give a brief discussion of the groups which can be obtained from the Lorentz transformations.

The Lorentz transformations form a group called the (homogeneous) Lorentz group. It is a subgroup of the *inhomogeneous* Lorentz group, also known as the *Poincaré* group. The latter group is formed from the inhomogeneous Lorentz transformations

$$x'^{\mu} = \Lambda^{\mu}{}_{\nu}x^{\nu} + x_0^{\mu}, \qquad (3.47)$$

where x_0^{μ} describes *translations*.

The Lorentz group possesses four disconnected parts which arise as follows.

(1) $L_+^{\uparrow}$: $\det\Lambda = +1$, $\Lambda^{0}{}_{0} \geq +1$. This part contains the identity element of the group. The aggregate of all proper, orthochronous, Lorentz trsansformations provides a group, which is a subgroup of the Lorentz group. It

is called the *proper, orthochronous, Lorentz group*.

(2) $L_-^\uparrow$: $\det \Lambda = -1$, $\Lambda^0{}_0 \geq +1$. This part contains a *space inversion* element S which describes a reflection relative to the three spatial axes:

$$\begin{aligned} x'^0 &= x^0, \\ x'^1 &= -x^1, \\ x'^2 &= -x^2, \\ x'^3 &= -x^3. \end{aligned} \qquad (3.48)$$

(3) $L_-^\downarrow$: $\det \Lambda = -1$, $\Lambda^0{}_0 \leq -1$. This part contains a *time reversal* element T which describes a reflection relative to the time axis:

$$\begin{aligned} x'^0 &= -x^0, \\ x'^1 &= x^1, \\ x'^2 &= x^2, \\ x'^3 &= x^3. \end{aligned} \qquad (3.49)$$

(4) $L_+^\downarrow$: $\det \Lambda = +1$, $\Lambda^0{}_0 \leq -1$. This part contains the element ST.

As was mentioned before, from the above four parts of the Lorentz group one obtaines the subgroup $L^\uparrow = L_+^\uparrow \bigcup L_-^\uparrow$ (the union of $L_+^\uparrow$ and $L_-^\uparrow$), called the *orthochronous Lorentz group*. Likewise, the subgroup $L_+ = L_+^\uparrow \bigcup L_+^\downarrow$, called the *proper Lorentz group*, is obtained.

Finally, we notice that every improper Lorentz transformation can be written in the form

$$\Lambda = S\Lambda_p, \qquad (3.50)$$

where S is a space-inversion element and Λ_p is a proper Lorentz transformation.

3.3 The Infinitesimal Approach

3.3.1 Infinitesimal Lorentz Matrices

Rotations $a_1(\psi)$, $a_2(\psi)$, $a_3(\psi)$ and Lorentz transformations (*boosts*) $b_1(\psi)$, $b_2(\psi)$, $b_3(\psi)$, around and along Ox_1, Ox_2, Ox_3 can then be written explicitly. These matrices are given by

$$a_1(\psi) = \begin{pmatrix} 1 & 0 & 0 & 0 \\ 0 & 1 & 0 & 0 \\ 0 & 0 & \cos\psi & -\sin\psi \\ 0 & 0 & \sin\psi & \cos\psi \end{pmatrix}, \qquad (3.51a)$$

$$a_2(\psi) = \begin{pmatrix} 1 & 0 & 0 & 0 \\ 0 & \cos\psi & 0 & \sin\psi \\ 0 & 0 & 1 & 0 \\ 0 & -\sin\psi & 0 & \cos\psi \end{pmatrix}, \qquad (3.51b)$$

$$a_3(\psi) = \begin{pmatrix} 1 & 0 & 0 & 0 \\ 0 & \cos\psi & -\sin\psi & 0 \\ 0 & \sin\psi & \cos\psi & 0 \\ 0 & 0 & 0 & 1 \end{pmatrix}, \qquad (3.51c)$$

and

$$b_1(\psi) = \begin{pmatrix} \cosh\psi & \sinh\psi & 0 & 0 \\ \sinh\psi & \cosh\psi & 0 & 0 \\ 0 & 0 & 1 & 0 \\ 0 & 0 & 0 & 1 \end{pmatrix}, \qquad (3.52a)$$

$$b_2(\psi) = \begin{pmatrix} \cosh\psi & 0 & \sinh\psi & 0 \\ 0 & 1 & 0 & 0 \\ \sinh\psi & 0 & \cosh\psi & 0 \\ 0 & 0 & 0 & 1 \end{pmatrix}, \qquad (3.52b)$$

$$b_3(\psi) = \begin{pmatrix} \cosh\psi & 0 & 0 & \sinh\psi \\ 0 & 1 & 0 & 0 \\ 0 & 0 & 1 & 0 \\ \sinh\psi & 0 & 0 & \cosh\psi \end{pmatrix}. \qquad (3.52c)$$

Infinitesimal Matrices

As in the case of the rotation group, the infinitesimal matrices a_r and b_r of the group L are defined by

$$a_r = \left[\frac{da_r(\psi)}{d\psi}\right]_{\psi=0}, \quad b_r = \left[\frac{db_r(\psi)}{d\psi}\right]_{\psi=0}. \qquad (3.53)$$

The a_r and b_r are related to $a_r(\psi)$ and $b_r(\psi)$ by

$$a_r(\psi) = \exp(\psi a_r), \quad b_r(\psi) = \exp(\psi b_r), \qquad (3.54)$$

and are given by

$$a_1 = \begin{pmatrix} 0 & 0 & 0 & 0 \\ 0 & 0 & 0 & 0 \\ 0 & 0 & 0 & -1 \\ 0 & 0 & 1 & 0 \end{pmatrix}, \qquad (3.55a)$$

3.3. THE INFINITESIMAL APPROACH

$$a_2 = \begin{pmatrix} 0 & 0 & 0 & 0 \\ 0 & 0 & 0 & 1 \\ 0 & 0 & 0 & 0 \\ 0 & -1 & 0 & 0 \end{pmatrix}, \qquad (3.55b)$$

$$a_3 = \begin{pmatrix} 0 & 0 & 0 & 0 \\ 0 & 0 & -1 & 0 \\ 0 & 1 & 0 & 0 \\ 0 & 0 & 0 & 0 \end{pmatrix}, \qquad (3.55c)$$

and

$$b_1 = \begin{pmatrix} 0 & 1 & 0 & 0 \\ 1 & 0 & 0 & 0 \\ 0 & 0 & 0 & 0 \\ 0 & 0 & 0 & 0 \end{pmatrix}, \qquad (3.56a)$$

$$b_2 = \begin{pmatrix} 0 & 0 & 1 & 0 \\ 0 & 0 & 0 & 0 \\ 1 & 0 & 0 & 0 \\ 0 & 0 & 0 & 0 \end{pmatrix}, \qquad (3.56b)$$

$$b_3 = \begin{pmatrix} 0 & 0 & 0 & 1 \\ 0 & 0 & 0 & 0 \\ 0 & 0 & 0 & 0 \\ 1 & 0 & 0 & 0 \end{pmatrix}. \qquad (3.56c)$$

One can easily see that these matrices satisfy the commutation relations

$$[a_i, \ a_j] = \epsilon_{ijk} a_k, \qquad (3.57a)$$

$$[b_i, \ b_j] = -\epsilon_{ijk} a_k, \qquad (3.57b)$$

$$[a_i, \ b_j] = \epsilon_{ijk} b_k. \qquad (3.57c)$$

Here ϵ_{ijk} is the usual Levi-Civita symbol defined by $\epsilon_{123} = 1$.

3.3.2 Infinitesimal Operators

We denote an arbitrary linear representation of the group L in an infinite-dimensional space B (see Chapter 2) by $g \to D(g)$ and for convenience we denote

$$A_r(\psi) = D(a_r(\psi)), \ B_r(\psi) = D(b_r(\psi)). \qquad (3.58)$$

$A_r(\psi)$ and $B_r(\psi)$ are continuous functions of ψ and are called basic one-parameter groups of operators for the given representation. They satisfy the relations (no summation on r)

$$A_r(\psi_1) A_r(\psi_2) = A_r(\psi_1 + \psi_2), \qquad (3.59a)$$

$$B_r(\psi_1) B_r(\psi_2) = B_r(\psi_1 + \psi_2). \qquad (3.59b)$$

$$A_r(0) = 1, \; B_r(0) = 1. \qquad (3.59c)$$

If the representation is finite-dimensional then the operators $A_r(\psi)$ and $B_r(\psi)$ are differentiable functions of ψ. If the representation is infinite-dimensional, however, these operators might be non-differentiable.

Basic Infinitesimal Operators

The basic infinitesimal operators of the one-parameter groups $A_r(\psi)$ and $B_r(\psi)$ are then defined by

$$A_r = \left[\frac{dA_r(\psi)}{d\psi}\right]_{\psi=0}, \; B_r = \left[\frac{dB_r(\psi)}{d\psi}\right]_{\psi=0}, \qquad (3.60)$$

if the representation is finite-dimensional. $A_r(\psi)$ and $B_r(\psi)$ might then be expanded in terms of A_r and B_r as

$$A_r(\psi) = \exp(\psi A_r), \; B_r(\psi) = \exp(\psi B_r). \qquad (3.61)$$

If the representation $g \to D(g)$ is infinite-dimensional, however, the operator functions $A_r(\psi)$ and $B_r(\psi)$ might be non-differentiable, but there may still exist a vector x for which $A_r(\psi) x$ and $B_r(\psi) x$ are differentiable vector-functions.

In general, let $A(t)$ be a continuous one-parameter group of operators in a Banach space R (see Chapter 2), and denote by $X(A)$ the set of all vectors $x \in R$ for which the limit of $(A(t) x - x)/t$, when $t \to 0$, exists in the sense of the norm in R. Obviously the set $X(A)$ contains the vector $x = 0$. Define now the operator A for all $x \in X(A)$ by $Ax = \lim\left[(A(t) x - x)/t\right]$ at the limit $t \to 0$. The domain of definition, $X(A)$, of the operator A is a subspace of R, and A is linear, i.e.,

$$A(\lambda_1 x_1 + \lambda_2 x_2) = \lambda_1 A x_1 + \lambda_2 A x_2 \text{ for } x_1, x_2 \in X(A). \qquad (3.62)$$

Such an operator A is called the infinitesimal operator of the one-parameter group $A(t)$. If $A(t) = D(a(t))$ is the group of operators of the representation $g \to D(g)$, corresponding to a one-parameter subgroup $a(t)$ of the group L, the corresponding operator A is then called the *infinitesimal operator* of the representation $g \to D(g)$.

3.3. THE INFINITESIMAL APPROACH

3.3.3 Determination of the Representation by its Infinitesimal Operators

A representation $g \to D(g)$ of the group L is completely determined by its infinitesimal operators A_i and B_i, $i = 1, 2, 3$. The determination of the irreducible representations of the group L is based on the fact that the basic infinitesimal operators of a representation satisfy the same commutation relations that exist among the infinitesimal matrices a_r and b_r, namely:

$$[A_i, A_j] = \epsilon_{ijk} A_k, \qquad (3.63a)$$

$$[B_i, B_j] = -\epsilon_{ijk} A_k, \qquad (3.63b)$$

$$[A_i, B_j] = \epsilon_{ijk} B_k. \qquad (3.63c)$$

Defining now the new infinitesimal operators

$$L_\mp = iA_1 \pm A_2, \quad L_3 = iA_3, \qquad (3.64a)$$

$$F_\mp = iB_1 \pm B_2, \quad F_3 = iB_3, \qquad (3.64b)$$

one then finds that they satisfy the following commutation relations:

$$[L_\mp, L_3] = \pm L_\mp, \quad [L_+, L_-] = 2L_3, \qquad (3.65a)$$

$$[F_\mp, F_3] = \mp L_\mp, \quad [F_+, F_-] = -2L_3, \qquad (3.65b)$$

$$[L_\pm, F_\pm] = 0, \quad [L_3, F_3] = 0, \qquad (3.65c)$$

$$[L_\pm, F_3] = \mp F_\pm, \quad [F_\pm, L_3] = \mp F_\pm, \qquad (3.65d)$$

$$[L_\pm, F_\mp] = \pm 2F_3. \qquad (3.65e)$$

The problem of determining a representation then reduces to the determination of $L_\pm$, L_3, $F_\pm$, F_3 satisfying the conditions (3.65).

Now, since the three-dimensional pure rotation group SO(3) is a subgroup of the proper, orthochronous Lorentz group L, obviously every representation of the group L is also a representation of the group SO(3). In fact, any infinite-dimensional representation of the group L, when regarded as a representation of the group SO(3), is highly reducible; it is equivalent to a direct sum of an infinite number of irreducible representations. The space of representation R of any irreducible representation of the group L is, therefore, a closed direct sum of subspaces M^j, where M^j is the $(2j+1)$-dimensional space in which the irreducible representation of weight j of the group SO(3) is realized.

Following the standard convention, one chooses the $2j+1$ normalized eigenvectors of the operator L_3 as the canonical basis for the subspace M^j. Let these base vectors be denoted as f_m^j, where $m = -j, -j+1, ..., j$, the superscript j indicates the subspace to which f_m^j belongs, and the subscript is the eigenvalue of the operator L_3. The superscript in f_m^j specifies the subspace uniquely since each irreducible representation of SO(3) is contained at most once in any given irreducible representation of the group L.

3.3.4 Conclusions

A detailed investigation of the commutation relations (3.65) in terms of the canonical basis f_m^j then leads to the following conclusions:

(a) Each irreducible representation of the group L is characterized by a pair of numbers (j_0, c), where j_0 is integral or half-integral, and c is a complex number.

(b) The space $R(j_0, c)$ of any given irreducible infinite-dimensional representation of the group L is characterized by integer or half-integer j_0 such that
$$R(j_0, c) = M^{j_0} \oplus M^{j_0+1} \oplus \cdots. \tag{3.66}$$
The whole space $R(j_0, c)$ is apanned, therefore, by the set of base vectors f_m^j, where $j = j_0, j_0+1, j_0+2, ...$, and $m = -j, -j+1, ..., j$. If the given irreducible representation is finite-dimensional then the direct sum of the subspaces M's terminates after a finite number of terms.

(c) A given reppresentation is finite-dimensional if and only if
$$c^2 = (j_0 + n)^2, \tag{3.67}$$
for some natural number n.

(d) The irreducible representation corresponding to a given pair (j_0, c) is, with a suitable choice of basis f_m^j in the space of representation, given by the formulas

$$L_\pm f_m^j = [(j \pm m + 1)(j \mp m)]^{1/2} f_{m\pm 1}^j, \tag{3.68a}$$

$$L_3 f_m^j = m f_m^j, \tag{3.68b}$$

$$\begin{aligned}F_\pm f_m^j = &\pm [(j \mp m)(j \mp m - 1)]^{1/2} C_j f_{m\pm 1}^{j-1} \\ &- [(j \pm m)(j \pm m + 1)]^{1/2} A_j f_{m\pm 1}^j \\ &\pm [(j \pm m + 1)(j \pm m + 2)]^{1/2} C_{j+1} f_{m\pm 1}^{j+1},\end{aligned} \tag{3.68c}$$

$$\begin{aligned}F_3 f_m^j = &[(j-m)(j+m)]^{1/2} C_j f_m^{j-1} - m A_j f_m^j \\ &- [(j+m+1)(j-m+1)]^{1/2} C_{j+1} f_m^{j+1}.\end{aligned} \tag{3.68d}$$

3.3. THE INFINITESIMAL APPROACH

Here we have used the symbols A_j and C_j which are defined by

$$A_j = \frac{icj_0}{j(j+1)}, \qquad (3.69a)$$

and

$$C_j = \frac{i\left(j^2 - j_0^2\right)^{1/2}\left(j^2 - c^2\right)^{1/2}}{j(4j^2 - 1)^{1/2}}. \qquad (3.69b).$$

(e) To each pair of numbers (j_0, c), where j_0 is integral or half-integral and c is complex, there corresponds a representation $g \to D(g)$ of the group L, whose infinitesimal operators are given by Eqs. (3.68).

Equations (3.68), for the unitary representations case and under certain assumptions, were first obtained by Gelfand (see the book of Naimark); they later on were rederived by Harish-Chandra, and by I.M. Gelfand and A.M. Iaglom.

3.3.5 Unitarity Conditions

A representation of a group G in a space R is called unitary if R is a Hilbert space (see Chapter 2) and $D(g)$ is a unitary operator for all $g \in G$. This implies that $(D(g)x, D(g)y) = (x, y)$ for all $g \in G$ and all $x, y \in R$, where (x, y) denotes the scalar product in R. (For the physical significance of non-unitary representations see Barut et al..) If the representation $g \to D(g)$ of the group L is unitary, then Eqs. (3.68) satisfy certain conditions which are summarized below.

Adjoint Operator

An operator B is called an adjoint to the operator A if $(Ax, y) = (x, By)$ for all $x, y \in R$. Let A be an infinitesimal operator of a unitary representation $g \to D(g)$ of the group L. Then $A(t) = D(A(t))$ is a unitary operator and therefore its adjoint satisfies:

$$[A(t)]^\dagger = A(-t). \qquad (3.70)$$

Accordingly one has

$$(A(t)f, g) = (f, A(-t)g). \qquad (3.71)$$

Differentiating both sides of this equation with respect to t we obtain for $t = 0$,

$$(Af, g) = -(f, Ag). \qquad (3.72)$$

Using this relation one then easily finds that

$$(L_+ f, g) = (f, L_- g), \qquad (3.73a)$$

$$(L_3 f, g) = (f, L_3 g), \qquad (3.73b)$$

$$(F_+ f, g) = (f, F_- g), \qquad (3.73c)$$

$$(F_3 f, g) = (f, F_3 g). \qquad (3.73d)$$

A systematic use of Eqs (3.73) in (3.68) then leads to the following.

If the irreducible representation $g \to D(g)$ of the group L is unitary then the pair (j_0, c) characterizing it satisfies either: (a) c is purely imaginary and j_0 is an arbitrary non-negative integral or half-integral number; or (b) c is a real number in the intervals $0 < |c| \le 1$ and $j_0 = 0$.

The representations corresponding to case (a) are called the *principal series of representations* and those corresponding to case (b) are called the *complementary series*.

3.4 The Group SL(2,C) and the Lorentz Group

We now introduce the group of all 2×2 complex matrices with determinant unity, the group SL(2,C), and establish a homomorphism between the group SL(2,C) and the proper, orthochronous, Lorentz group (see Subsection 3.2.2). Subgroups of the group SL(2,C) are then discussed, and the connection with the Lobachevskian motion is pointed out.

3.4.1 The Group SL(2,C)

In what follows we establish the fact that elements of the proper, orthochronous, homogeneous Lorentz group L discussed in the last section can be described by means of elements of SL(2,C), the group of all 2×2 complex matrices

$$g = \begin{pmatrix} a & b \\ c & d \end{pmatrix} \qquad (3.74)$$

with

$$\det g = ad - bc = 1. \qquad (3.75)$$

In the natural topology of matrices the group SL(2,C) is simply connected. The relation between the two groups can be established as follows.

3.4. THE GROUP SL(2,C) AND THE LORENTZ GROUP

One associates with each four-vector x^μ a Hermitian matrix

$$Q = \begin{pmatrix} x^0 + x^3 & x^1 + ix^2 \\ x^1 - ix^2 & x^0 - x^3 \end{pmatrix}. \tag{3.76}$$

In this way one defines a one-to-one linear correspondence between all four-vectors and all 2×2 Hermitian matrices. Equation (3.76) can also be written as

$$Q = x_\alpha \sigma^\alpha, \tag{3.77}$$

where σ^k, $k = 1, 2, 3$, are the three Pauli matrices and σ^0 is the 2×2 unit matrix:

$$\sigma^0 = \begin{pmatrix} 1 & 0 \\ 0 & 1 \end{pmatrix}, \sigma^1 = \begin{pmatrix} 0 & 1 \\ 1 & 0 \end{pmatrix}, \sigma^2 = \begin{pmatrix} 0 & i \\ -i & 0 \end{pmatrix}, \sigma^3 = \begin{pmatrix} 1 & 0 \\ 0 & -1 \end{pmatrix}. \tag{3.78}$$

It is often also very convenient to parametrize the elements g of the group SL(2,C) by

$$g = g_\mu \sigma^\mu, \tag{3.79}$$

where g_0, g_k, $k = 1, 2, 3$, are complex numbers.

Corresponding to every element g of the group SL(2,C) consider the following transformation in the space of the Hermitian matrices Q:

$$Q' = gQg^\dagger, \tag{3.80}$$

where $g^\dagger$ is the Hermitian conjugate of g, and $Q' = x'_\alpha \sigma^\alpha$. The corresponding operation in the Minkowskian space of four-vectors is a linear transformation

$$x'^\alpha = \Lambda^\alpha{}_\beta(g) x^\beta, \tag{3.81a}$$

or, in matrix notation,

$$x' = \Lambda(g) x, \tag{3.81b}$$

where the transformation matrix Λ can be expressed in terms of the matrix g of the group SL(2,C). The transformation (3.81) preserves the scalar product since

$$(x'^0)^2 - (x'^1)^2 - (x'^2)^2 - (x'^3)^2 = \det Q'$$

$$= \det Q = (x^0)^2 - (x^1)^2 - (x^2)^2 - (x^3)^2. \tag{3.82}$$

3.4.2 Homomorphism of the Group SL(2,C) on the Lorentz Group L

The matrix elements $\Lambda^\alpha{}_\beta$ can be expressed in terms of the corresponding matrix g of the group SL(2,C). Using the properties of the Pauli spin matrices, and using Eq. (3.80) and Eqs. (3.81), one has

$$x'^\alpha = \delta^\alpha_\beta x'^\beta = \frac{1}{2}\text{Tr}\left(\sigma^\alpha \sigma^\beta\right) x'_\beta = \frac{1}{2}\text{Tr}\left(\sigma^\alpha Q'\right)$$

$$= \frac{1}{2}\text{Tr}\left(\sigma^\alpha g Q g^\dagger\right) = \frac{1}{2}\text{Tr}\left(\sigma^\alpha g \sigma^\beta g^\dagger\right) x_\beta. \tag{3.83}$$

Comparing this result with Eq. (3.81a) one obtains

$$\Lambda^{\alpha\beta} = \frac{1}{2}\text{Tr}\left(\sigma^\alpha g \sigma^\beta g^\dagger\right), \tag{3.84a}$$

where $g^\dagger$ is the Hermitian conjugate of the matrix g, and Tr stands for trace.

The explicit expression of the transformation $\Lambda(g)$ in terms of the parameters g_0 and g_k of the matrix g is as follows (Problem 3.1):

$$\Lambda^0{}_0 = |g_0|^2 + \sum_{k=1}^{3} |g_k|^2, \tag{3.84b}$$

$$\Lambda^k{}_0 = g_0 \bar{g}_k + \bar{g}_0 g_k - i\epsilon^{klm} g_l \bar{g}_m, \tag{3.84c}$$

$$\Lambda^0{}_k = g_0 \bar{g}_k + \bar{g}_0 g_k + i\epsilon^{klm} g_l \bar{g}_m, \tag{3.84d}$$

$$\Lambda^l{}_k = \delta^l_k \left(|g_0|^2 - \sum_{s=1}^{3} |g_s|^2\right) + g_k \bar{g}_l + \bar{g}_k g_l - i\epsilon^{klm}\left(\bar{g}_0 g_m - g_0 \bar{g}_m\right), \tag{3.84e}$$

where the ϵ symbols are fixed by $\epsilon^{123} = \epsilon^{0123} = +1$. In particular, one notices the useful relation

$$\text{Tr}\,\Lambda(g) = |\text{Tr}\,g|^2 = 4|g_0|^2. \tag{3.85}$$

One also notices that because the group SL(2,C) is connected, and the mapping into the homogeneous Lorentz group is a continuous homomorphism, the image of the group SL(2,C) must be a subgroup of the proper orthochronous, Lorentz group L.

3.4. THE GROUP SL(2,C) AND THE LORENTZ GROUP

Equations (3.84) show that to an arbitrary matrix g of SL(2,C) there corresponds a 4×4 matrix Λ. We now show that the matrix Λ belongs to the proper, orthochronous, Lorentz group L.

First, from Eq. (3.82), one sees the quadratic form $(x^0)^2 - (x^1)^2 - (x^2)^2 - (x^3)^2$ is invariant under the transformation Λ, and therefore the matrix Λ is an element of the homogeneous Lorentz group. As a consequence, $\det \Lambda = \pm 1$. But for the special case for which g is the 2×2 unit matrix, the corresponding Λ is the identity transformation, and hence $\det \Lambda = 1$. Since $\det \Lambda$ is a continuous function of the four variables a, b, c, d of the matrix g of the group SL(2,C), and since the domain of variation of these four variables is simply connected, a discontinuous jump from $\det \Lambda = +1$ to $\det \Lambda = -1$ is excluded.

Consequently, $\det \Lambda = +1$ for all values of a, b, c, d, subject to the restriction (3.75). Hence Λ belongs to the proper Lorentz group. Finally, from Eqs. (3.84) one sees that $\Lambda^0{}_0$ cannot be negative. Accordingly, Λ is orthochronous. Consequently, Λ is an element of the proper, orthochronous, Lorentz group L.

Suppose now that an element Λ of the group L is given. Let us try to invert the relations (3.84). If $\operatorname{Tr} \Lambda \neq 0$ we obtain (Problem 3.2):

$$g = g_0 \sigma^0 + \sum_{k=1}^{3} g_k \sigma^k$$

$$= D^{-1} \left[\operatorname{Tr} \Lambda \sigma^0 + \sum_{k=1}^{3} \left(\Lambda^k{}_0 + \Lambda^0{}_k - i\epsilon^{0k\rho}_{\sigma} \Lambda^\sigma{}_\rho \right) \sigma^k \right], \qquad (3.86)$$

where

$$D^2 = 4 - \operatorname{Tr} \Lambda^2 + (\operatorname{Tr} \Lambda)^2 - i\epsilon^{\mu\lambda}_{\rho\sigma} \Lambda^\sigma{}_\lambda \Lambda^\rho{}_\mu. \qquad (3.87)$$

The sign of the denominator D is undetermined. Since the smallest subgroup of the group L that contains all elements with $\operatorname{Tr} \Lambda \neq 0$ is Λ itself, the image of the group SL(2,C) is the whole of the group L.

It is possible to find the elements g of SL(2,C) which go into L in the case $\operatorname{Tr} \Lambda = 0$, also. If

$$\sum_{k=1}^{3} \left(\Lambda^k{}_0 \right)^2 \neq 0, \qquad (3.88)$$

the matrix Λ then describes a rotation with an angle π, and one has

$$g_0 = 0, \quad g = \sum_{k=1}^{3} g_k \sigma^k, \quad g^2 = -e, \qquad (3.89a)$$

$$\Lambda(g)\Lambda(g) = \Lambda(g^2) = \Lambda(-e) = I, \qquad (3.89b)$$

where e is the 2×2 unit matrix. The 3×3 matrix $M^l_{\ k} = \delta^l_{\ k}\Lambda^0_{\ 0} + \Lambda^l_{\ k}$ is symmetric and possesses the three eigenvalues 0, $\Lambda^0_{\ 0} - 1$, $\Lambda^0_{\ 0} + 1$. The eigenvalue 0 belongs to the normalized eigenvector $\mathbf{v}_0$ defined by

$$(v_0)^k = \left[\sum_{l=1}^{3} (\Lambda^l_{\ 0})^2\right]^{-1/2} \Lambda^k_{\ 0}. \qquad (3.90)$$

If we denote the normalized real eigenvector, corresponding to the eigenvalue $\Lambda^0_{\ 0} + 1$, by $\mathbf{v}_1$, we can then express the components g_k in terms of the vectors $\mathbf{v}_1$ and $\mathbf{v}_1 \times \mathbf{v}_0$, as follows:

$$g_k = \pm\left[\left(\frac{1}{2}(\Lambda^0_{\ 0} - 1)\right)^{1/2}(\mathbf{v}_1 \times \mathbf{v}_0)^k + i\left(\frac{1}{2}(\Lambda^0_{\ 0} + 1)\right)^{1/2} v_1^k\right]. \qquad (3.91)$$

Again the matrix g is determined only up to a sign. The remaining case, for which $\text{Tr } \Lambda = \Lambda^k_{\ 0} = 0$, $k = 1, 2, 3$, is contained in Eq. (3.91) as the limit $\Lambda^0_{\ 0} = 1$. One obtains (Problem 3.3):

$$g_k = \pm i v_1^k. \qquad (3.92)$$

In this fashion one reaches the conclusion that there exists a two-to-one mapping between all the elements of the group SL(2,C) and all the elements Λ of the proper, orthochronous, Lorentz group L such that to each element Λ of the group L there correspond two elements $\pm g$ of the group SL(2,C), and to each element g of the group SL(2,C) there corresponds an element Λ of the group L. The mapping preserves the group multiplication and constitutes, therefore, a homomorphism of the group SL(2,C) on the group L. As a result of this, the description of the representations of the group L is equivalent to that of the group SL(2,C); a representation $g \to D(g)$ of L is single- or double-valued according to whether or not $D(g)$ is equal to $D(-g)$ or not.

3.4.3 Kernel of Homomorphism

The sign ambiguity of $g = g(\Lambda)$ means, in particular, that the unit matrix I of the group L is the image of both central elements $e_\pm$ of the group SL(2,C), where $e_\pm = \pm e$, and e is the 2×2 unit matrix. (Group elements are called central if they commute with all group elements. Central elements form the center of the group.) Hence we have established an isomorphism between

3.4. THE GROUP SL(2,C) AND THE LORENTZ GROUP

the proper, orthochronous, Lorentz group L and the group SL(2,C)/Z_2, where Z_2 denotes the center of the group SL(2,C) consisting of the elements $e_\pm$.

3.4.4 Subgroups of the Group SL(2,C)

The group SL(2,C) possesses some important subgroups, some of which play crucial roles in further investigations. These subgroups correspond to subgroups of the proper, orthochronous, homogeneous, Lorentz group L as well. Since the group SL(2,C) is more natural to handle than the group L, one prefers to deal with the group SL(2,C)×T_4, where T_4 is the translational group. The group SL(2,C)×T_4 is sometimes called the *inhomogeneous SL(2,C) group*. (See Subsection 3.2.2 for the inhomogeneous Lorentz group.)

The group SU(2) has already been mentioned as a subgroup of SL(2,C). It consists, of course, of those elements u satisfying $u^\dagger = u^{-1}$. A possible parametrization of the group SU(2) is as follows:

$$u = u_\alpha \sigma^\alpha, \tag{3.93}$$

with the condition

$$u_0^2 + \sum_{k=1}^{3} u_k^2 = 1. \tag{3.94}$$

Here u_0 and u_k, $k = 1, 2, 3$, are real numbers.

Another subgroup of SL(2,C) is the group SU(1,1). It consists of those elements v of SL(2,C) satisfying the condition $v^\dagger \sigma^3 v = \sigma^3$. A possible parametrization is as follows:

$$v = v_0 \sigma^0 + v_1 \sigma^1 + v_2 \sigma^2 + i v_3 \sigma^3, \tag{3.95}$$

with the condition

$$v_0^2 - v_1^2 - v_2^2 + v_3^2 = 1. \tag{3.96}$$

Here the numbers v_0 and v_k, $k = 1, 2, 3$, are real.

A third subgroup of SL(2,C) is the group SL(2,R). It consists of elements a of the group SL(2,C) satisfying $a^\dagger \sigma^2 a = \sigma^2$. They can be presented as

$$a = a_0 \sigma^0 + a_1 \sigma^1 + i a_2 \sigma^2 + a_3 \sigma^3. \tag{3.97}$$

Here the numbers a_0 and a_k, $k = 1, 2, 3$, are real and satisfy the condition

$$a_0^2 - a_1^2 + a_2^2 - a_3^2 = 1. \tag{3.98}$$

The matrix a is a real 2×2 matrix.

By the rotation $\exp(i\pi\sigma^1/4) = 2^{-1/2}(\sigma^0 + i\sigma^1)$ in the $x^2 - x^3$ plane, we can map the group SU(1,1) on the group SL(2,R):

$$a = \exp\left(-\frac{i\pi\sigma^1}{4}\right) v \exp\left(\frac{i\pi\sigma^1}{4}\right), \qquad (3.99)$$

from which one infers that $a_0 = v_0$, $a_1 = v_1$, $a_2 = v_3$, and $a_3 = -v_2$. This one-to-one mapping of the two groups SU(1,1) and SL(2,R) onto each other is sometimes called the *standard isomorphism*.

Finally, the group of triangular matrices

$$(\phi, \mu) = \begin{pmatrix} e^{-i\phi/2} & 0 \\ \mu e^{-i\phi/2} & e^{i\phi/2} \end{pmatrix}, \qquad (3.100)$$

where μ is complex and $0 \leq \phi \leq 4\pi$, with the group multiplication law $(\phi_1, \mu_1) \times (\phi_2, \mu_2) = (\phi_1 + \phi_2 (\pm 4\pi), \mu_1 + e^{i\phi_1}\mu_2)$, is isomorphic to the group of Euclidean motions on the Riemannian plane of functions $z^{1/2}$. The corresponding subgroup of L is isomorphic to the group of motions in the complex z-plane itself. The notation of this subgroup is U(1)×T$_2$ if one means the subgroup of L, and U(1)′×T$_2$ if one means the subgroup of SL(2,C).

3.4.5 Connection with Lobachevskian Motions

We have seen that each complex, unimodular, two-dimensional matrix g induces a Lorentz transformation in the Minkowskian space according to $Q' = gQg^\dagger$, where Q is given by Eq. (3.76). These Lorentz transformations map the surfaces

$$x_0^2 - x_1^2 - x_2^2 - x_3^2 = c \qquad (3.101)$$

into themselves, since they preserve the corresponding quadratic form.

There are three types of such surfaces. These are either sheet of a two-sheeted hyperboloid when $c > 0$, a single-sheeted hyperboloid when $c < 0$, and either the positive or the negative cone when $c = 0$. [If, instead of considering points x in Minkowskian space, we deal with Hermitian matrices Q, the surfaces would be the following three types of manifolds in the space of Hermitian matrices: all positive definite (or negative definite) Hermitian matrices with fixed determinant $c > 0$; all Hermitian matrices with fixed determinant $c < 0$; and all Hermitian matrices $Q \geq 0$ (or $Q \leq 0$), that is matrices Q whose corresponding Hermitian form takes on nonnegative (or

3.5. PROBLEMS

nonpositive) values, with determinant zero. See, for example, Gelfand *et al.*]

The Lorentz transformations induce transformations that are called *motions* of these surfaces. In this way to each complex unimodular two-dimensional matrix g there corresponds a motion on each of the surfaces above. One can show that a given motion corresponds to two matrices g_1 and g_2 if and only if $g_1 = \pm g_2$.

The upper sheet of a two-sheeted hyperboloid together with the motions defined in this way is one model of *Lobachevskian space*. This means that the *group of complex two-dimensional unimodular matrices is locally isomorphic to the group of Lobachevskian motions*. In addition to Lobachevskian space, there exist two related spaces with groups of motions locally isomorphic to the same group of matrices. Models of these spaces are the single-sheeted hyperboloid and the positive cone.

We conclude this brief discussion on the Lobachevskian space by pointing out that the group of motions on each of these surfaces is *transitive*, that is every point of the space can be transformed by some motion to any other point. Let us prove this assertion for the upper sheet of the two-sheeted hyperboloid $x_0^2 - x_1^2 - x_2^2 - x_3^2 = 1$, as the proof for the other surfaces is similar. Using Eq. (3.76), then the points on our surface correspond to positive definite *unimodular* Hermitian matrices. Since every such matrix can be written in the form $Q = gg^\dagger = geg^\dagger$, where g is a complex unimodular matrix and e is the 2×2 unit matrix. This proves that there exists a motion transforming the fixed unit matrix into Q.

In the next chapter the theory of two-component spinors is developed in detail.

3.5 Problems

3.1 Use Eq. (3.84a) to prove Eqs. (3.84b)-(3.84e).

Solution: The solution is left for the reader.

3.2 Prove Eq. (3.86).

Solution: The solution is left for the reader. (See the book of Rühl.)
3.3 Prove Eq. (3.92).

Solution: The solution is left for the reader. (See Rühl.)

3.4 Prove that the kernel of the homomorphism of the group SL(2,C) onto the group L coincides with the center of the group SL(2,C). (Notice that by definition the element g is in the kernel of the homomorphism if for all Hermitian matrices Q one has $Q = gQg^\dagger$.)

Solution: The solution is left for the reader.

3.5 Use Eq. (3.84a) to show that if a, b, c and d are the four elements of the matrix g of the group SL(2,C), with $ad - bc = 1$, then the corresponding matrix Λ of the proper, orthochronous, homogeneous, Lorentz group is given by

$$\Lambda = \begin{pmatrix} \frac{1}{2}\left(a\bar{a}+b\bar{b}\right) \\ +\frac{1}{2}\left(c\bar{c}+d\bar{d}\right) & \Re\left(a\bar{b}+c\bar{d}\right) & \Im\left(\bar{a}b-c\bar{d}\right) & \frac{1}{2}\left(a\bar{a}-b\bar{b}\right) \\ +\frac{1}{2}\left(c\bar{c}-d\bar{d}\right) \\ \Re\left(a\bar{c}+b\bar{d}\right) & \Re\left(a\bar{d}+b\bar{c}\right) & \Im\left(\bar{a}d+b\bar{c}\right) & \Re\left(a\bar{c}-b\bar{d}\right) \\ \Im\left(a\bar{c}+b\bar{d}\right) & \Im\left(a\bar{d}+b\bar{c}\right) & \Re\left(\bar{a}d-b\bar{c}\right) & \Im\left(a\bar{c}-b\bar{d}\right) \\ \frac{1}{2}\left(a\bar{a}+b\bar{b}\right) \\ -\frac{1}{2}\left(c\bar{c}+d\bar{d}\right) & \Re\left(a\bar{b}-c\bar{d}\right) & \Im\left(\bar{a}b+c\bar{d}\right) & \frac{1}{2}\left(a\bar{a}-b\bar{b}\right) \\ -\frac{1}{2}\left(c\bar{c}-d\bar{d}\right) \end{pmatrix},$$

where $\Re$ and $\Im$ denote real and imaginary parts. Show that the same matrix can also be obtained directly from either Eq. (3.80) or Eqs. (3.84b)-(3.84e).

Solution: The solution is left for the reader.

3.6 References for Further Reading

A.O. Barut, H. Kleinert and S. Malin, The "anomalous Zitterbewegung" of composite particles, *Nuovo Cimento* **58A**, 835-847 (1968). (Section 3.3)

A.O. Barut and S. Malin, Position operators and localizability of quantum systems described by finite- and infinite-dimensional wave equations, *Revs. Mod. Phys.* **40**, 632-651 (1968). (Section 3.3)

D. Bohm, *The Special Theory of Relativity* (Benjamin, New York, 1965). (Section 3.1)

M. Born, *Einstein's Theory of Relativity* (Dover, New York, 1962). (Section 3.1)

M. Carmeli, *Group Theory and General Relativity* (McGraw-Hill, New York, 1977).

M. Carmeli, *Cosmological Special Relativity: The Large-Scale Structure of Space, Time and Velocity* (World Scientific, Singapore, 1997). (Section 3.1)

M. Carmeli and S. Malin, Finite- and infinite-dimensional representations

3.6. REFERENCES FOR FURTHER READING

of the Lorentz group, *Fortschritte der Physik* **21**, 397-425 (1973). (Section 3.3)

M. Carmeli and S. Malin, *Representations of the Rotation and Lorentz Groups* (Marcel Dekker, New York and Basel, 1976).

A. Einstein, Zur Elektrodynamik bewegter Körper, *Ann. Physik* **17**, 891-921 (1905); English translation, On the electrodynamics of moving bodies, in: *The Principle of Relativity* (Dover, New York, 1923). (Section 3.1)

A.P. French, *Special Relativity* (W.W. Norton, New York and London, 1968). (Section 3.1)

I.M. Gelfand, M.I. Graev and N.Ya. Vilenkin, *Integral Geometry and Representation Theory* (Academic Press, New York, 1966). (Section 3.4)

I.M. Gelfand and A.M. Iaglom, General relativistic invariant equations and finite-dimensional representations of the Lorentz group, *Zh. Eksp. Theor. Fiz.* **18**, 703 (1948). (Section 3.3)

Harish-Chandra, Infinite irreducible representations of the Lorentz group, *Proc. Roy. Soc. (London) A* **189**, 372 (1947); On relativistic wave equations, *Phys. Rev.* **71**, 793 (1947). (Section 3.3)

A.I. Miller, *Albert Einstein's Special Theory of Relativity* (Addison-Wesley, Reading, Massachusets, 1981). (Section 3.1)

M.A. Naimark, *Linear Representations of the Lorentz Group* (Pergamon Press, New York, 1964). (Section 3.3)

W. Rühl, *The Lorentz Group and Harmonic Analysis* (W.A. Benjamin, New York, 1970). (Sections 3.2, 3.4)

Chapter 4

Two-Component Spinors

Spinors were invented by Elie Cartan without reference to the theory of representations of groups. The natural way to study spinors, however, is through the representation theory of the groups SU(2) and SL(2,C); two-component spinors occur in the spinor representation of the group SL(2,C). The latter is an irreducible, finite-dimensional, nonunitary representation. It can be shown that any finite-dimensional representation of the group SL(2,C) is equivalent to the spinor representation. In this chapter we study the above-mentioned topics. Infinite-dimensional spinors are subsequently given.

4.1 Spinor Representation of SL(2,C)

We now construct the spinor representation which contains all the finite-dimensional irreducible representations of the group SL(2,C). A generalization to infinite dimensions is given in the last section.

4.1.1 The Space of Polynomials

Let us denote by P_{mn} the aggregate of all polynomials with complex coefficients $p(z, \overline{z})$ of the two variables z and $\overline{z}$, where z is a complex variable and $\overline{z}$ is its complex conjugate. We assume that the polynomial $p(z, \overline{z})$ is of degree not exceeding m in the variable z and not exceeding n in the variable $\overline{z}$, where m and n are two fixed nonnegative integers. Hence we have

$$p(z,\bar{z}) = \sum_{r,s=0}^{m,n} p_{rs} z^r \bar{z}^s = p_{00} + p_{10}z + p_{01}\bar{z} + \cdots + p_{mn}z^m\bar{z}^n. \quad (4.1)$$

The space P_{mn} is therefore determined by the two integers m and n.

The space P_{mn} may be considered a linear vector space, the components of the "vectors" being the coefficients p_{rs}, with $r = 0, 1, \ldots, m$ and $s = 0, 1, \ldots, n$. For each value of s there are $m+1$ values for r, and for each value of r there are $n+1$ values for s. Hence the dimension of the space P_{mn} is $(m+1)(n+1)$. The operation of addition of two polynomials in P_{mn} is defined, as usual, by

$$p'(z,\bar{z}) + p''(z,\bar{z}) = \sum p'_{rs} z^r \bar{z}^s + \sum p''_{rs} z^r \bar{z}^s = \sum p_{rs} z^r \bar{z}^s, \quad (4.2)$$

where $p_{rs} = p'_{rs} + p''_{rs}$.

The operation of product by a number is also defined, as usual, by

$$ap(z,\bar{z}) = a \sum p_{rs} z^r \bar{z}^s = \sum p'_{rs} z^r \bar{z}^s, \quad (4.3)$$

where $p'_{rs} = a p_{rs}$. We obviously have

$$ap(z,\bar{z}) + bp(z,\bar{z}) = (a+b)p(z,\bar{z}), \quad (4.4)$$

$$ap'(z,\bar{z}) + ap''(z,\bar{z}) = a(p'+p'')(z,\bar{z}). \quad (4.5)$$

Thus P_{mn} is a linear space. The space P_{mn} will be used in the following as the space of representation for the group SL(2,C).

4.1.2 Realization of the Spinor Representation

Let us now denote an element of the group SL(2,C) by

$$g = \begin{pmatrix} a & b \\ c & d \end{pmatrix}, \quad ad - bc = 1, \quad (4.6)$$

where a, b, c and d are four complex numbers. We then define the operator $D(g)$ in the space P_{mn} by

$$D(g)p(z,\bar{z}) = (bz+d)^m (\bar{b}\bar{z}+\bar{d})^n p(w,\bar{w}), \quad (4.7)$$

where w is the image of z under the Mobius transformation

$$w = \frac{az+c}{bz+d}. \quad (4.8)$$

4.1. SPINOR REPRESENTATION OF SL(2,C)

Accordingly, to each element g of the group SL(2,C) there corresponds an operator $D(g)$ defined in the linear space P_{mn}.

We now verify that the correspondence $g \to D(g)$ is a linear representation of the group SL(2,C). To this end we have to show that

$$D(g_1) D(g_2) p(z,\bar{z}) = D(g_1 g_2) p(z,\bar{z}), \quad (4.9)$$

$$D(I) = 1, \quad (4.10)$$

for arbitrary elements g_1 and g_2 of the group SL(2,C). In the above formulas I denotes the unity element of SL(2,C) and 1 denotes the unit operator in the space P_{mn}.

Let now the elements g_1 and g_2 of the group SL(2,C) be denoted by

$$g_1 = \begin{pmatrix} a_1 & b_1 \\ c_1 & d_1 \end{pmatrix}, \quad g_2 = \begin{pmatrix} a_2 & b_2 \\ c_2 & d_2 \end{pmatrix}, \quad (4.11)$$

and hence their product is given by

$$g = g_1 g_2 = \begin{pmatrix} a_1 a_2 + b_1 c_2 & a_1 b_2 + b_1 d_2 \\ c_1 a_2 + d_1 c_2 & c_1 b_2 + d_1 d_2 \end{pmatrix} = \begin{pmatrix} a & b \\ c & d \end{pmatrix}. \quad (4.12)$$

Accordingly we have, using the representation formula (4.7),

$$D(g_2) p(z,\bar{z}) = (b_2 z + d_2)^m (\bar{b}_2 \bar{z} + \bar{d}_2)^n p(w_2, \bar{w}_2), \quad (4.13)$$

where w_2 is the image of z under the Mobius transformation associated with the matrix g_2,

$$w_2 = \frac{a_2 z + c_2}{b_2 z + d_2}. \quad (4.14)$$

Applying now the operator $D(g_1)$ on both sides of Eq. (4.13), we obtain

$$D(g_1) D(g_2) p(z,\bar{z})$$
$$= (b_1 z + d_1)^m (\bar{b}_1 \bar{z} + \bar{d}_1)^n (b_2 w_1 + d_2)^m (\bar{b}_2 \bar{w}_1 + \bar{d}_2)^n p(v, \bar{v}), \quad (4.15)$$

where w_1 is the image of z under the Mobius transformation corresponding to the element g_1 of SL(2,C),

$$w_1 = \frac{a_1 z + c_1}{b_1 z + d_1}, \quad (4.16)$$

and v is obtained from w_2 by replacing the variable z by w_1,

$$v = \frac{a_2 w_1 + c_2}{b_2 w_1 + d_2}. \quad (4.17)$$

A simple calculation then gives the following for the products of the terms with equal powers m and n in Eq. (4.15):

$$(b_1 z + d_1)(b_2 w_1 + d_2) = (a_1 b_2 + b_1 d_2) z + (c_1 b_2 + d_1 d_2) = bz + d. \quad (4.18)$$

Accordingly we obtain

$$D(g_1) D(g_2) p(z, \bar{z}) = (bz + d)^m \left(\overline{bz} + \overline{d}\right)^n p(v, \bar{v}) \quad (4.19)$$

for Eq.(4.15).

The variable v of Eq. (4.17) may be calculated using Eqs.(4.16) and (4.12), giving

$$v = \frac{(a_1 a_2 + b_1 c_2) z + (c_1 a_2 + d_1 c_2)}{(a_1 b_2 + b_1 d_2) z + (c_1 b_2 + d_1 d_2)} = \frac{az + c}{bz + d}. \quad (4.20)$$

In the above formulas a, b, c and d are the elements of the matrix $g = g_1 g_2$ given by Eq. (4.12). Accordingly we obtain for Eq. (4.19), using Eq. (4.20), the following:

$$D(g_1) D(g_2) p(z, \bar{z}) = D(g_1 g_2) p(z, \bar{z}), \quad (4.21)$$

thus proving Eq. (4.9). The proof of Eq. (4.10) is immediate since $D(I) p(z, \bar{z}) = p(z, \bar{z})$ by the representation formula (4.7).

The correspondence $g \to D(g)$ is thus a finite-dimensional representation of the group SL(2,C) since it is being realized in the finite-dimensional space P_{mn}. It is known as the *spinor representation* and is usually denoted by $D^{\left(\frac{m}{2}, \frac{n}{2}\right)}$. Its dimension is, of course, equal to $(m+1)(n+1)$.

4.1.3 Two-Component Spinors

In order to introduce the two-component spinors we realize the spinor representation $D^{\left(\frac{m}{2}, \frac{n}{2}\right)}$ discussed above in a somewhat different form.

To this end let us consider all systems of numbers

$$\phi_{A_1 \cdots A_m X'_1 \cdots X'_n}, \quad (4.22)$$

which are symmetric in their indices $A_1 \cdots A_m$ and $X'_1 \cdots X'_n$, taking the values 0, 1 and 0', 1', respectively. Such numbers may be considered as the components of vectors of a linear space. Let us denote such a space by $\tilde{P}_{mn}$. Because of the symmetry of the indices in (4.22), we actually have only $m + 1$ independent indices $A_1 \cdots A_m$ and $n + 1$ independent indices

4.1. SPINOR REPRESENTATION OF SL(2,C)

$X'_1 \cdots X'_n$. These are $0, \ldots, 0, 0; 0, \ldots, 0, 1; \ldots; 1, \ldots, 1, 1$, for instance, for $A_1 \cdots A_{m-1} A_m$. Hence the dimension of the space $\tilde{P}_{mn}$ is equal to $(m+1)(n+1)$.

We may relate the two spaces P_{mn} and $\tilde{P}_{mn}$ by one-to-one mapping by associating to each number (4.22) of $\tilde{P}_{mn}$ the polynomial

$$p(z, \bar{z}) = \sum \phi_{A_1 \cdots A_m X'_1 \cdots X'_n} z^{A_1 + \cdots + A_m} \bar{z}^{X'_1 + \cdots + X'_n}. \tag{4.23}$$

This polynomial is of degree not larger than m in the variable z and not larger than n in the variable $\bar{z}$. Hence the polynomial (4.23) belongs to the space P_{mn}.

On the other hand every polynomial

$$p(z, \bar{z}) = \sum p_{rs} z^r \bar{z}^s \tag{4.24}$$

of the space P_{mn} may be written in the form given by Eq. (4.23) by relating the coefficients of $z^r \bar{z}^s$ of the two polynomials (4.23) and (4.24). We then obtain

$$\binom{m}{r} \binom{n}{s} \phi_{A_1 \cdots A_m X'_1 \cdots X'_n} = p_{rs} \tag{4.25}$$

along with the conditions

$$A_1 + \cdots + A_m = r, \quad X'_1 + \cdots + X'_n = s. \tag{4.26}$$

In Eq. (4.25)

$$\binom{m}{n} = \frac{m!}{(m-n)! n!}. \tag{4.27}$$

A second form of the spinor representation is obtained if we apply the operator $D(g)$ on the polynomial (4.23). We then obtain

$$D(g) p(z, \bar{z}) = D(g) \sum \phi_{B_1 \cdots B_m Y'_1 \cdots Y'_n} z^{B_1 + \cdots + B_m} \bar{z}^{Y'_1 + \cdots + Y'_n}$$

$$= (g_1{}^0 z + g_0{}^0)^m (\bar{g}_1{}^0 \bar{z} + \bar{g}_0{}^0)^n \sum \phi_{B_1 \cdots B_m Y'_1 \cdots Y'_n} w^{B_1 + \cdots + B_m} \bar{w}^{Y'_1 + \cdots + Y'_n}, \tag{4.28}$$

where w is given by Eq. (4.8) and use has been made of the notation $g_1{}^1 = a$, $g_1{}^0 = b$, $g_0{}^1 = c$, and $g_0{}^0 = d$. Hence we obtain

$$D(g) p(z, \bar{z}) = \sum \phi_{B_1 \cdots B_m Y'_1 \cdots Y'_n} (g_1{}^1 z + g_0{}^1)^{B_1 + \cdots + B_m}$$

$$\times (g_1{}^0 z + g_0{}^0)^{m - B_1 - \cdots - B_m} (\bar{g}_{1'}{}^{1'} \bar{z} + \bar{g}_{0'}{}^{1'})^{Y'_1 + \cdots + Y'_n}$$

70   CHAPTER 4. TWO-COMPONENT SPINORS

$$\times \left(\bar{g}_{1'}{}^{0'} \bar{z} + \bar{g}_{0'}{}^{0'} \right)^{n - Y_1' - \cdots - Y_n'}, \tag{4.29}$$

and therefore

$$D(g)p(z,\bar{z}) = \sum g_0{}^{B_1} \cdots g_0{}^{B_m} \bar{g}_{0'}{}^{Y_1'} \cdots \bar{g}_{0'}{}^{Y_n'} \phi_{B_1 \cdots B_m Y_1' \cdots Y_n'}$$

$$+ \cdots + g_1{}^{B_1} \cdots g_1{}^{B_m} \bar{g}_{1'}{}^{Y_1'} \cdots \bar{g}_{1'}{}^{Y_n'} \phi_{B_1 \cdots B_m Y_1' \cdots Y_n'} z^m \bar{z}^n$$

$$= \sum_{A,X'} \left(\sum_{B,Y'} g_{A_1}{}^{B_1} \cdots g_{A_m}{}^{B_m} \bar{g}_{X_1'}{}^{Y_1'} \cdots \bar{g}_{X_n'}{}^{Y_n'} \phi_{B_1 \cdots B_m Y_1' \cdots Y_n'} \right)$$

$$\times z^{A_1 + \cdots + A_m} \bar{z}^{X_1' + \cdots + X_n'}. \tag{4.30}$$

Accordingly we may finally write the following for the spinor representation:

$$D(g)p(z,\bar{z}) = \sum \phi'_{A_1 \cdots A_m X_1' \cdots X_n'} z^{A_1 + \cdots + A_m} \bar{z}^{X_1' + \cdots + X_n'}, \tag{4.31}$$

where

$$\phi'_{A_1 \cdots A_m X_1' \cdots X_n'} = \sum_{B,Y'} g_{A_1}{}^{B_1} \cdots g_{A_m}{}^{B_m} \bar{g}_{X_1'}{}^{Y_1'} \cdots \bar{g}_{X_n'}{}^{Y_n'} \phi_{B_1 \cdots B_m Y_1' \cdots Y_n'} \tag{4.32}$$

is the transformed ϕ under the group SL(2,C).

The quantities $\phi_{A_1 \cdots A_m X_1' \cdots X_n'}$ are called two-component spinors, and are complex numbers. The indices $A_1 \cdots A_m$ take the values 0, 1 and are called unprimed (or undotted) indices, whereas $X_1' \cdots X_n'$ take the values $0'$, $1'$ and are called primed (or dotted) indices. Similarly to tensors, every spinor has an order. The spinor ϕ defined above, for instance, is of order m in its unprimed indices and n in its primed indices. Equation (4.32) shows that these two kinds of indices transform under elements of the group SL(2,C) and its complex conjugate, respectively. The summation over the indices can also be made to run over 0, 1 instead of over 1, 0 by relabeling the matrix g of SL(2,C) so that $g_1{}^1 = a$, $g_1{}^0 = b$, $g_0{}^1 = c$, and $g_0{}^0 = d$, where a, b, c and d are defined by Eq. (4.6).

Finally we notice that although two-component spinors were introduced above as numbers, they can actually be made functions of spacetime when applied in physics. This is again similar to tensors. The essential difference between tensors and spinors is their association with groups. While tensors are associated with the Lorentz group, spinors are associated with the group SL(2,C), which is the covering group of the Lorentz group. As a result of

4.1. SPINOR REPRESENTATION OF SL(2,C)

this fact, spinors can be used to describe particles with spins $\frac{1}{2}$, $\frac{3}{2}$,... in addition to those with spins 0, 1, 2,..., whereas tensors can describe only the latter kind of particles. As a consequence, spinors are considered to be more fundamental than tensors from both the mathematical and the physical points of view.

4.1.4 Examples

1. The spinor representation $D^{(\frac{1}{2},\frac{1}{2})}$.

The representation $D^{(\frac{1}{2},\frac{1}{2})}$ corresponds to $m = 2j_1 = 1$ and $n = 2j_2 = 1$. The space of representation has accordingly the dimension of $(2j_1 + 1)(2j_2 + 1) = (m+1)(n+1) = 4$. The space P_{mn} is the aggregate of all polynomials of the form

$$p(z, \bar{z}) = p_{00} + p_{10}z + p_{01}\bar{z} + p_{11}z\bar{z}. \tag{1}$$

When the operator $D(g)$ is applied to the above polynomial, we obtain

$$D(g) p(z, \bar{z}) = (bz + d)(\bar{b}\bar{z} + \bar{d})(p_{00} + p_{10}w + p_{01}\bar{w} + p_{11}w\bar{w})$$

$$= p_{00}(bz + d)(\bar{b}\bar{z} + \bar{d}) + p_{10}(az + c)(\bar{b}\bar{z} + \bar{d})$$
$$+ p_{01}(bz + d)(\bar{a}\bar{z} + \bar{c}) + p_{11}(az + c)(\bar{a}\bar{z} + \bar{c}), \tag{2}$$

by the representation formulas (4.7) and (4.8).

Using now the correspondence between the spaces P_{mn} and $\tilde{P}_{mn}$ we then find the following, using Eq. (4.25), for the relationship between the polynomial coefficients p_{rs} and the components of the corresponding spinor:

$$p_{00} = \phi_{00'}, \quad p_{10} = \phi_{10'}, \quad p_{01} = \phi_{01'}, \quad p_{11} = \phi_{11'}, \tag{3}$$

Hence we have a spinor with two indices, one is unprimed and one is primed.

The polynomial $p(z, \bar{z})$ is therefore defined in the space $\tilde{P}_{mn}$ by

$$p(z, \bar{z}) = \sum \phi_{AX'} z^A \bar{z}^{X'} = \phi_{00'} + \phi_{10'}z + \phi_{01'}\bar{z} + \phi_{11'}z\bar{z}. \tag{4}$$

Using now Eqs. (2) and (3) we obtain

$$D(g) p(z, \bar{z}) = \phi_{00'} \left(g_1{}^0 z + g_0{}^0\right) \left(\bar{g}_{1'}{}^{0'} \bar{z} + \bar{g}_{0'}{}^{0'}\right)$$

$$+ \phi_{10'} \left(g_1{}^1 z + g_0{}^1\right) \left(\bar{g}_{1'}{}^{0'} \bar{z} + \bar{g}_{0'}{}^{0'}\right)$$

$$+\phi_{01'}\left(g_1{}^0z+g_0{}^0\right)\left(\bar{g}_{1'}^{1'}\bar{z}+\bar{g}_{0'}^{1'}\right)$$
$$+\phi_{11'}\left(g_1{}^1z+g_0{}^1\right)\left(\bar{g}_{1'}^{1'}\bar{z}+\bar{g}_{0'}^{1'}\right), \tag{5}$$

which may also be written in the form

$$D(g)p(z,\bar{z})=\phi_{00'}\left(g_1{}^0\bar{g}_{1'}^{0'}z\bar{z}+\cdots+g_0{}^0\bar{g}_{0'}^{0'}\right)+\cdots$$
$$+\phi_{11'}\left(g_1{}^1\bar{g}_{1'}^{1'}z\bar{z}+\cdots+g_0{}^1\bar{g}_{0'}^{1'}\right). \tag{6}$$

Hence we have

$$D(g)p(z,\bar{z})=\left(g_0{}^0\bar{g}_{0'}^{0'}\phi_{00'}+\cdots+g_0{}^1\bar{g}_{0'}^{1'}\phi_{11'}\right)+\cdots$$
$$+\left(g_1{}^0\bar{g}_{1'}^{0'}\phi_{00'}+\cdots+g_1{}^1\bar{g}_{1'}^{1'}\phi_{11'}\right)z\bar{z}$$
$$=\sum g_0{}^B\bar{g}_{0'}^{Y'}\phi_{BY'}+\cdots+\sum g_1{}^B\bar{g}_{1'}^{Y'}\phi_{BY'}z\bar{z}, \tag{7}$$

or

$$D(g)p(z,\bar{z})=\sum_{A,X'}\left(\sum_{B,Y'}g_A{}^B\bar{g}_{X'}^{Y'}\phi_{BY'}\right)z^A\bar{z}^{X'}$$
$$=\sum_{A,X'}\phi'_{AX'}z^A\bar{z}^{X'}, \tag{8}$$

where

$$\phi'_{AX'}=\sum_{B,Y'}g_A{}^B\bar{g}_{X'}^{Y'}\phi_{BY'}. \tag{9}$$

In the above formulas $A, B = 1, 0$ and $X', Y' = 1', 0'$ (or, alternatively, 0,1 and 0',1').

2. The spinor corresponding to the space of representation with dimension $m = 2j_1 = 2$ and $n = 2j_2 = 1$.

The polynomials of the space P_{21} are given by

$$p(z,\bar{z})=p_{00}+p_{10}z+p_{01}\bar{z}+p_{11}z\bar{z}+p_{20}z^2+p_{21}z^2\bar{z}. \tag{1}$$

By Eq. (4.25) the spinor corresponding to the coefficients p_{rs} is given by $\phi_{ABX'}$, namely, of order 2 in its unprimed indices and of order 1 in its primed indices. One then easily finds that

$$\begin{aligned}p_{00}&=\phi_{000'}, & p_{10}&=2\phi_{010'}=2\phi_{100'},\\ p_{01}&=\phi_{001'}, & p_{11}&=2\phi_{011'}=2\phi_{101'},\\ p_{20}&=\phi_{110'}, & p_{21}&=\phi_{111'}.\end{aligned} \tag{2}$$

4.2. OPERATORS OF THE SPINOR REPRESENTATION

Hence we finally have for the polynomial $p(z,\bar{z})$, in terms of the spinor $\phi_{ABX'}$ the following:

$$p(z,\bar{z}) = \phi_{000'} + (\phi_{010'} + \phi_{100'})z + \phi_{001'}\bar{z}$$
$$+ (\phi_{011'} + \phi_{101'})z\bar{z} + \phi_{110'}z^2\phi_{111'}z^2\bar{z}$$
$$= \sum \phi_{ABX'} z^{A+B} \bar{z}^{X'}. \tag{3}$$

4.2 Operators of the Spinor Representation

We now find the infinitesimal operators $L_\mp$, L_3, $K_\mp$, K_3 of the spinor representation for the group SL(2,C) discussed in the last section.

4.2.1 One-Parameter Subgroups

First we find the one-parameter subgroups of the group SL(2,C) corresponding to the one-parameter subgroups $a_k(\psi)$ and $b_k(\psi)$ of the proper, orthochronous, homogeneous Lorentz group L. These can easily be derived using Eqs. (3.86). One finds for these one-parameter groups:

$$a_1(\psi) = \begin{pmatrix} \cos\frac{\psi}{2} & i\sin\frac{\psi}{2} \\ i\sin\frac{\psi}{2} & \cos\frac{\psi}{2} \end{pmatrix}, \quad a_2(\psi) = \begin{pmatrix} \cos\frac{\psi}{2} & -\sin\frac{\psi}{2} \\ \sin\frac{\psi}{2} & \cos\frac{\psi}{2} \end{pmatrix},$$

$$a_3(\psi) = \begin{pmatrix} e^{i\psi/2} & 0 \\ 0 & e^{-i\psi/2} \end{pmatrix}, \tag{4.33a}$$

$$b_1(\psi) = \begin{pmatrix} \cosh\frac{\psi}{2} & \sinh\frac{\psi}{2} \\ \sinh\frac{\psi}{2} & \cosh\frac{\psi}{2} \end{pmatrix}, \quad b_2(\psi) = \begin{pmatrix} \cosh\frac{\psi}{2} & i\sinh\frac{\psi}{2} \\ -i\sinh\frac{\psi}{2} & \cosh\frac{\psi}{2} \end{pmatrix},$$

$$b_3(\psi) = \begin{pmatrix} e^{\psi/2} & 0 \\ 0 & e^{-\psi/2} \end{pmatrix}, \tag{4.33b}$$

In terms of the infinitesimal matrices a_k and b_k, where $k = 1, 2, 3$, of the group SL(2,C), they can be written as

$$a_k(\psi) = e^{\psi a_k}, \quad b_k(\psi) = e^{\psi b_k}, \tag{4.34}$$

where a_k and b_k are given in terms of the three Pauli matrices (3.78) by $a_k = i\sigma^k/2$ and $b_k = \sigma^k/2$.

4.2.2 Infinitesimal Operators

Using now Eqs. (4.7) and (4.8) one can find the operators $A_k(\psi)$ and $B_k(\psi)$, where $k = 1, 2, 3$. For example,

$$A_1(\psi)\, p(z, \bar{z}) = \left(i \sin \frac{\psi}{2} z + \cos \frac{\psi}{2}\right)^m \left(-i \sin \frac{\psi}{2} \bar{z} + \cos \frac{\psi}{2}\right)^n$$

$$\times p \left(\frac{\cos \frac{\psi}{2} z + i \sin \frac{\psi}{2}}{i \sin \frac{\psi}{2} z + \cos \frac{\psi}{2}},\ \frac{\cos \frac{\psi}{2} \bar{z} - i \sin \frac{\psi}{2}}{-i \sin \frac{\psi}{2} \bar{z} + \cos \frac{\psi}{2}} \right), \qquad (4.35a)$$

$$A_2(\psi)\, p(z, \bar{z}) = \left(-\sin \frac{\psi}{2} z + \cos \frac{\psi}{2}\right)^m \left(-\sin \frac{\psi}{2} \bar{z} + \cos \frac{\psi}{2}\right)^n$$

$$\times p \left(\frac{\cos \frac{\psi}{2} z + \sin \frac{\psi}{2}}{-\sin \frac{\psi}{2} z + \cos \frac{\psi}{2}},\ \frac{\cos \frac{\psi}{2} \bar{z} + \sin \frac{\psi}{2}}{-\sin \frac{\psi}{2} \bar{z} + \cos \frac{\psi}{2}} \right), \qquad (4.35b)$$

$$A_3(\psi)\, p(z, \bar{z}) = e^{-im\psi/2} e^{in\psi/2} p\left(e^{i\psi} z, e^{-i\psi} \bar{z}\right). \qquad (4.35c)$$

In the same way one finds the operators $B_k(\psi)$, $k = 1, 2, 3$. Differentiating both sides of these equations with respect to the parameter ψ, and putting $\psi = 0$, one obtains:

$$A_1 p = \left[\frac{i}{2}(1 - z^2) \frac{\partial}{\partial z} - \frac{i}{2}(1 - \bar{z}^2) \frac{\partial}{\partial \bar{z}} + \frac{i}{2}(mz - n\bar{z})\right] p, \qquad (4.36a)$$

$$A_2 p = \left[\frac{1}{2}(1 + z^2) \frac{\partial}{\partial z} + \frac{1}{2}(1 + \bar{z}^2) \frac{\partial}{\partial \bar{z}} - \frac{1}{2}(mz + n\bar{z})\right] p, \qquad (4.36b)$$

$$A_3 p = \left[iz \frac{\partial}{\partial z} - i\bar{z} \frac{\partial}{\partial \bar{z}} - \frac{i}{2}(m - n)\right] p, \qquad (4.36c)$$

$$B_1 p = \left[\frac{1}{2}(1 - z^2) \frac{\partial}{\partial z} + \frac{1}{2}(1 - \bar{z}^2) \frac{\partial}{\partial \bar{z}} + \frac{1}{2}(mz + n\bar{z})\right] p, \qquad (4.36d)$$

$$B_2 p = \left[-\frac{i}{2}(1 + z^2) \frac{\partial}{\partial z} + \frac{i}{2}(1 + \bar{z}^2) \frac{\partial}{\partial \bar{z}} + \frac{i}{2}(mz - n\bar{z})\right] p, \qquad (4.36e)$$

$$B_3 p = \left[z \frac{\partial}{\partial z} + \bar{z} \frac{\partial}{\partial \bar{z}} - \frac{1}{2}(m + n)\right] p, \qquad (4.36f)$$

4.2. OPERATORS OF THE SPINOR REPRESENTATION

The operators $L_\pm$, L_3, $K_\pm$, and K_3 can now be found, using Eqs. (4.36):

$$L_+ p = \left[-\frac{\partial}{\partial z} - \bar{z}^2 \frac{\partial}{\partial \bar{z}} + n\bar{z}\right] p, \tag{4.37a}$$

$$L_- p = \left[z^2 \frac{\partial}{\partial z} + \frac{\partial}{\partial \bar{z}} - mz\right] p, \tag{4.37b}$$

$$L_3 p = \left[-z \frac{\partial}{\partial z} + \bar{z}\frac{\partial}{\partial \bar{z}} + \frac{1}{2}(m-n)\right] p, \tag{4.37c}$$

$$K_+ p = \left[i\frac{\partial}{\partial z} - i\bar{z}^2 \frac{\partial}{\partial \bar{z}} + in\bar{z}\right] p, \tag{4.37d}$$

$$K_- p = \left[-iz^2 \frac{\partial}{\partial z} + i\frac{\partial}{\partial \bar{z}} + imz\right] p, \tag{4.37e}$$

$$K_3 p = \left[iz\frac{\partial}{\partial z} + i\bar{z}\frac{\partial}{\partial \bar{z}} - \frac{i}{2}(m+n)\right] p, \tag{4.37f}$$

4.2.3 Matrix Elements of the Spinor Operator $D(g)$

To conclude this section we find the matrices of the spinor operators $D(g)$. For more details see the book of Rühl.

Consider the complex two-vectors $\xi = (\xi^1, \xi^2)$, which transform as

$$\xi' = \xi g, \tag{4.38}$$

under application of a matrix g of the group SL(2,C). Let us construct a linear vector space of complex polynomials by defining $p(\xi) = p(\xi^1, \xi^2)$ which are homogeneous in ξ^1 and ξ^2 of degree $2J$, where $2J$ is an arbitrary nonnegative integer. In this space, having the dimension $(2J+1)$, we define the transformation $D(g)$ for any g of SL(2,C) by

$$D(g) p(\xi) = p(\xi') = p(\xi g). \tag{4.39}$$

These transformations provide a $(2J+1)$-dimensional representation of the group SL(2,C).

In order to relate the transformation (4.39) to a more familiar notation we expand the polynomial p into powers of ξ^1 and ξ^2:

$$p(\xi) = \sum_{M=-J}^{J} \chi_M^J N_M^J \left(\xi^1\right)^{J+M} \left(\xi^2\right)^{J-M}. \tag{4.40}$$

Here N_M^J are normalized constants defined by

$$N_M^J = \left[\frac{(2J)!}{(J+M)!\,(J-M)!}\right]^{1/2}, \tag{4.41}$$

and χ_M^J are expansion coefficients. In terms of χ_M^J the transformation $D(g)$ can be expressed as

$$\left(D(g)\chi^J\right)_M = \sum_{M'=-J}^{J} D_{MM'}^J(g)\,\chi_{M'}^J. \tag{4.42}$$

From Eqs. (4.39), (4.40), and (4.42) one finds

$$D_{MM'}^J(g) = \left[\frac{(J+M)!\,(J-M)!}{(J+M')!\,(J-M')!}\right]^{1/2} \sum_n \binom{J+M'}{n}\binom{J-M'}{J+M-n}$$

$$\times g_{11}^n g_{12}^{J+M-n} g_{21}^{J+M'-n} g_{22}^{n-M-M'}, \tag{4.43}$$

where the sum over n extends over all integers for which neither of the binomial coefficients vanish.

If we restrict the group SL(2,C) to the subgroup SU(2), we obtain the matrix $D_{MM'}^J(u)$ defining the unitary irreducible representation of the group SU(2). Each element $D_{MM'}^J(u)$ is a homogeneous polynomial of degree $2J$ in the matrix elements of u, and the coefficients of the polynomial are real. A substitution $u_{ij} \to g_{ij}$, with $i,j = 1,2$, leads us back to the matrix elements $D_{MM'}^J(g)$. If we substitute

$$\begin{aligned} u_{11} &= \alpha + i\beta, \\ u_{12} &= \gamma + i\delta, \\ u_{21} &= -\gamma + i\delta, \\ u_{22} &= \alpha - i\beta, \end{aligned} \tag{4.44}$$

with real α, β, γ, δ, such that

$$\alpha^2 + \beta^2 + \gamma^2 + \delta^2 = 1, \tag{4.45}$$

the substitution $u_{ij} \to g_{ij}$ can be interpreted as an extension of the real parameters α, β, γ, δ into the complex domain. This is called a *complexification* of the group SU(2).

4.2.4 Further Properties of Spinor Representations

We conclude this section by the following statements.

(1) *The spinor representations of the group SL(2,C) are irreducible.*

(2) *Every finite-dimensional irreducible representation of SL(2,C) is equivalent to some spinor representation. The pair of numbers j_0 and c, appearing in the representation formula, is then related to the pair of numbers m and n of the spinor representation by*

$$j_0 = \frac{1}{2} \mid m - n \mid, \tag{4.46a}$$

$$c = \begin{cases} [\text{sign}\,(m-n)]\left[\frac{1}{2}(m+n)+1\right]; & m \neq n \\ \pm\left[\frac{1}{2}(m+n)+1\right]; & m=n \end{cases} \tag{4.46b}$$

(3) *The spinor representations are all nonunitary.*

The proofs of (1), (2), and (3) are left for the reader (Problems 4.4-4.6).

4.3 Infinite-Dimensional Spinors

In this section the method used in Section 4.1 to obtain symmetrical spinors and their transformation law from finite-dimensional representations of the group SL(2,C) is extended to infinite-dimensional representations.

4.3.1 Principal Series of Representations

As we have seen in Section 4.1, two-component spinors are associated with finite-dimensional representations of the group SL(2,C) when realized in the space of polynomials. Spinors appear (up to factorial terms) as the coefficients of the polynomials of the space in which the representation is realized. Their transformation law then provides another form for the representation.

The group SL(2,C), however, has also infinite-dimensional representations the most notable of which is the *principal series of representations* (see Subsection 3.3.5). In this section we define an infinite set of numbers which can be associated with the principal series of representations in a way which is very similar, but as a generalization, to that two-component spinors appear in describing the finite-dimensional representations. The transformation law of these numbers, at the same time, defines another form of the principal series of representations of SL(2,C). Just as in the spinor case, these numbers become functions of spacetime when applied in physics.

The principal series of representations of SL(2,C) is an irreducible unitary representation, which can be realized in several ways according to the space of realization. For our purpose, we employ that particular realization of it by means of the special unitary group SU(2).

The Hilbert Space $L_2^{2s}(\mathrm{SU}(2))$

We denote by $L_2^{2s}(\mathrm{SU}(2))$ the set of all functions $\phi(u)$, where u is an element of SU(2), which are measurable and satisfy the conditions

$$\phi(\gamma u) = e^{is\psi}\phi(u), \qquad (4.47)$$

$$\int |\phi(u)|^2 \, du < \infty, \qquad (4.48)$$

where $\gamma \in \mathrm{SU}(2)$ is given by

$$\gamma = \begin{pmatrix} e^{-\frac{1}{2}i\psi} & 0 \\ 0 & e^{\frac{1}{2}i\psi} \end{pmatrix}. \qquad (4.49)$$

$L_2^{2s}(\mathrm{SU}(2))$ provides a Hilbert space where the scalar product is defined by

$$(\phi_1, \phi_2) = \int \phi_1(u) \overline{\phi_2}(u) \, du. \qquad (4.50)$$

The principal series of representations is then given by the formula

$$V(g)\phi(u) = \frac{\alpha(ug)}{\alpha(u\bar{g})}\phi(u\bar{g}), \qquad (4.51)$$

where

$$g = \begin{pmatrix} g_{11} & g_{12} \\ g_{21} & g_{22} \end{pmatrix}$$

is an element of the group SL(2,C) and $\alpha(g)$ is a function given by

$$\alpha(g) = g_{22}^{2s} |g_{22}|^{i\rho - 2s - 2}. \qquad (4.52)$$

Here ρ is a real number and $2s$ is an integer.

4.3. INFINITE-DIMENSIONAL SPINORS

4.3.2 Infinite-Dimensional Spinors

The Hilbert Space l_2^{2s}

Consider now all possible systems of numbers ϕ_m^j, where $m = -j, -j+1, \cdots, j$ and $j = |s|, |s|+1, |s|+2, \cdots$, with the condition

$$\sum_{j=|s|}^{\infty} (2j+1) \sum_{m=-j}^{j} |\phi_m^j|^2 < \infty. \tag{4.53}$$

The aggregate of all such systems ϕ_m^j forms a Hilbert space, which we denote by l_2^{2s}, where the scalar product is defined by

$$\sum_{j=|s|}^{\infty} (2j+1) \sum_{m=-j}^{j} \phi_m^j \overline{\psi_m^j}, \tag{4.54}$$

for any two vectors ϕ_m^j and ψ_m^k of l_2^{2s}.

With each vector $\phi_m^j \in l_2^{2s}$, we associate the function

$$\phi(u) = \sum_{j=|s|}^{\infty} (2j+1) \sum_{m=-j}^{j} \phi_m^j D_m^j(u), \tag{4.55}$$

where $D_m^j(u)$ is the matrix element $D_{sm}^j(u)$ of the irreducible representation of SU(2). Since

$$D_m^j(\gamma u) = e^{is\psi} D_m^j(u), \tag{4.56}$$

the function given by Eq. (4.55) belongs to the space $L_2^{2s}(\text{SU}(2))$.

On the other hand, every function in $L_2^{2s}(\text{SU}(2))$ can be written in the form (4.55), since the $D_m^j(u)$ provide a complete orthogonal set,

$$\int D_m^j(u) \overline{D_{m'}^{j'}}(u) \, du = \frac{1}{2j+1} \delta^{jj'} \delta_{mm'}. \tag{4.57}$$

The two spaces $L_2^{2s}(\text{SU}(2))$ and l_2^{2s} are, in fact, isometric where the transition from one space to the other can be made by means of the generalized Fourier transform

$$\phi_m^j = \int \phi(u) \overline{D_m^j}(u) \, du. \tag{4.58}$$

Similarly to spinors, which appear as coefficients in the polynomials of the space of representation, we see that the numbers ϕ_m^j appear as coefficients in the expansion given by Eq. (4.55) of the functions $\phi(u)$ of the space $L_2^{2s}(SU(2))$. By means of the mapping (4.58), the operator $V(g)$ of the representation (4.51) may also be regarded as an operator in the space l_2^{2s}, whose explicit expression we find below. This expression also defines another form of the principal series of representations.

Applying the operator $V(g)$ to the function $\phi(u)$ as given by Eq. (4.55), we obtain

$$V(g)\phi(u) = \sum_j (2j+1) \sum_m \phi_m^j \frac{\alpha(ug)}{\alpha(u\bar{g})} D_m^j(u\bar{g}), \qquad (4.59)$$

or

$$V(g)\phi(u) = \sum_j (2j+1) \sum_m \phi_m^j \sum_{j'} (2j'+1)$$

$$\times \sum_{m'} V_{mm'}^{jj'}(g;s,\rho) D_{m'}^{j'}(u), \qquad (4.60)$$

where

$$V_{mm'}^{jj'}(g;s,\rho) = \int \frac{\alpha(ug)}{\alpha(u\bar{g})} D_m^j(u\bar{g}) \overline{D}_{m'}^{j'}(u) \, du. \qquad (4.61)$$

Accordingly, we obtain

$$V(g)\phi(u) = \sum_j (2j+1) \sum_m \phi_m'^j D_m^j(u), \qquad (4.62)$$

where, using Eq. (14), we have

$$\phi_{m'}'^{j'} = \sum_{j=|s|}^{\infty} (2j+1) \sum_{m=-j}^{j} V_{mm'}^{jj'}(g;s,\rho) \phi_m^j. \qquad (4.63)$$

Thus, the operator $V(g)$ of the principal series of representations of SL(2,C) in the space l_2^{2s} is the linear transformation determined by Eq. (4.63) describing the law of transformation of the quantities ϕ_m^j, where $j = |s|, |s|+1, |s|+2, \cdots$ and $m = -j, -j+1, \cdots, j$. Here, $V_{mm'}^{jj'}(g;s,\rho)$ are functions of $g \in$ SL(2,C) and of ρ and s, where ρ is a real number and $2s$ is an integer.

In the next two chapters we apply the two-component spinors to curved spacetime topics.

4.4 Problems

4.1 Show that the matrix elements $D^J_{MM'}(g)$ of Eq. (4.43) satisfy the properties:
$$D^J_{MM'}(\bar{g}) = \overline{D}^J_{MM'}(g),$$
$$D^J_{MM'}(g^t) = D^J_{MM'}(g),$$
$$D^J_{MM'}\left((g^{-1})^t\right) = (-1)^{M-M'} D^J_{-M,-M'}(g).$$

Solution: The solution is left for the reader.

4.2 Show that the parameter ψ of Eqs. (4.33b) is related to the relative speed v of the Lorentz transformation by
$$\psi = \cosh^{-1}\left[(1 - v^2/c^2)^{-1/2}\right].$$

Solution: The solution is left for the reader.

4.3 Find the operators $B_k(\psi)$ for the spinor representation of the group SL(2,C) and from them prove Eqs. (4.36d)-(4.36f).

Solution: The solution is left for the reader.

4.4 Show that the spinor representations of the group SL(2,C) are irreducible.

Solution: The solution is left for the reader.

4.5 Prove Eqs. (4.46).

Solution: The solution is left for the reader.

4.6 Prove that the spinor representations are all nonunitary. [Thus the group SL(2,C) does not contain finite-dimensional unitary representations.]

Solution: The solution is left for the reader.

4.5 References for Further Reading

H. Boener, *Representations of Groups* (North-Holland, Amsterdam, The Netherlands, 1963). (Sections 4.1-4.3)

R. Brauer and H. Weyl, Spinors in n dimensions, *American J. Mathematics* **57**, 425 (1935). (Sections 4.1-4.3)

M. Carmeli, Infinite-dimensional representations of the Lorentz group, *J. Math. Phys.* **11**, 1917-1918 (1970). (Section 4.3)

M. Carmeli, *Classical Fields: General Relativity and Gauge Theory* (John Wiley, 1982).

M. Carmeli and S. Malin, Finite- and infinite-dimensional representations of the Lorentz group, *Fortschr. der Phys.* **21**, 397-425 (1973). (Sections 4.1, 4.3)

E. Cartan, *The Theory of Spinors* (The M.I.T. Press, Cambridge, Massachusetts, 1966). (Sections 4.1-4.3)

C. Chevalley, *The Algebraic Theory of Spinors* (Columbia University Press, New York, 1954). (Sections 4.1-4.3)

C. Chevalley, *Theory of Lie Groups* (Princeton University Press, Princeton, New Jersey, 1962). (Sections 4.1-4.3)

P.A.M. Dirac, *Spinors in Hilbert Space* (Plenum Press, New York, 1974). (Sections 4.1-4.3)

I.M. Gelfand, M.I. Graev and N.Ya. Vilenkin, *Generalized Functions, Vol.5: Integral Geometry and Representation Theory* (Academic Press, New York, 1966) (Sections 4.1, 4.3).

M.A. Naimark, *Linear Representations of the Lorentz Group* (Pergamon Press, New York, 1964). (Sections 4.1, 4.3)

W. Rühl, *The Lorentz Group and Harmonic Analysis* (Benjamin, New York, 1970). (Section 4.2)

B.L. van der Waerden, *Nachr. Wiss. Gottingen, Math.-Physik* **100** (1929). (Sections 4.1-4.3)

H. Weyl, *The Classical Groups* (Princeton University Press, Princeton, New Jersey, 1964). (Sections 4.1-4.3)

Chapter 5

Maxwell, Dirac and Pauli Spinors

In the last chapter we derived the two-component spinors and their transformation law under the group SL(2,C) from the theory of representations. We now apply the two-component spinors to curved spacetime. Hence these quantities will be functions of spacetime. In this chapter and the following chapters two-component spinors will be applied to the electromagnetic, Dirac, neutrino and gravitational fields. The first three fields can be introduced in flat (Minkowskian) spacetime, but not the gravitational field. Our presentation, however, will all be done in curved spacetime since that needs no further effort. It should be very easy to go from the curved to the Minkowskian spacetime if it is necessary. Hence our discussion will start with spinors in curved spacetime. We start the chapter with a brief review of Maxwell's theory.

5.1 Maxwell's Theory

In this section we give a brief account on the Maxwell equations for electrodynamics.

The Lagrangian density for the electromagnetic field is given by

$$L = -\frac{1}{16\pi}f_{\alpha\beta}f^{\alpha\beta} - \frac{1}{c}j^\alpha A_\alpha + L_e. \tag{5.1}$$

The field $f_{\mu\nu}$ is related to the potential A_α by

$$f_{\mu\nu} = \partial_\nu A_\mu - \partial_\mu A_\nu, \tag{5.2}$$

and j^α is the four-current density. L_e is the Lagrangian density of the charged particles.

Maxwell's equations are then given by

$$\partial_\nu f^{\mu\nu} = \frac{4\pi}{c} j^\mu, \tag{5.3}$$

$$\partial_\gamma f_{\alpha\beta} + \partial_\beta f_{\gamma\alpha} + \partial_\alpha f_{\beta\gamma} = 0. \tag{5.4}$$

Equation (5.3) is the field equation obtained from the Lagrangian density (5.1), whereas Eq. (5.4) is a consequence of Eq. (5.2).

The electric field **E** and the magnetic field **H** are related to the electromagnetic field tensor $f_{\mu\nu}$ by the following identification:

$$\mathbf{E} = (E_x, E_y, E_z) = (E_1, E_2, E_3), \tag{5.5a}$$

$$\mathbf{H} = (H_x, H_y, H_z) = (H_1, H_2, H_3), \tag{5.5b}$$

where E_i and H_i, with $i = 1, 2, 3$, are given by

$$E_i = f_{i0}, \qquad H_i = \frac{1}{2}\epsilon_{ijk} f_{jk}. \tag{5.6}$$

Here ϵ_{ijk} is the three-dimensional totally skew-symmetric tensor with values $+1$ and -1, depending upon whether ijk is an even or an odd permutation of 123, and zero otherwise.

The electromagnetic field tensors $f_{\mu\nu}$ and $f^{\mu\nu}$ may then be written explicitly as follows:

$$f_{\mu\nu} = \begin{pmatrix} 0 & -E_x & -E_y & -E_z \\ E_x & 0 & H_z & -H_y \\ E_y & -H_z & 0 & H_x \\ E_z & H_y & -H_x & 0 \end{pmatrix}, \tag{5.7a}$$

$$f^{\mu\nu} = \begin{pmatrix} 0 & E_x & E_y & E_z \\ -E_x & 0 & H_z & -H_y \\ -E_y & -H_z & 0 & H_x \\ -E_z & H_y & -H_x & 0 \end{pmatrix}. \tag{5.7b}$$

In terms of the dual $^\star f_{\mu\nu}$ to the electromagnetic field tensor $f_{\alpha\beta}$, given by

$$^\star f^{\alpha\beta} = \frac{1}{2}\epsilon^{\alpha\beta\mu\nu} f_{\mu\nu}, \tag{5.8}$$

5.1. MAXWELL'S THEORY

the Maxwell equation (5.4) may also be written in the alternative form

$$\frac{\partial *f^{\alpha\beta}}{\partial x^{\beta}} = 0. \tag{5.9}$$

We then have for the dual the following explicit expressions:

$$*f_{\mu\nu} = \begin{pmatrix} 0 & -H_x & -H_y & -H_z \\ H_x & 0 & -E_z & E_y \\ H_y & E_z & 0 & -E_x \\ H_z & -E_y & E_x & 0 \end{pmatrix}, \tag{5.10a}$$

$$*f^{\mu\nu} = \begin{pmatrix} 0 & H_x & H_y & H_z \\ -H_x & 0 & -E_z & E_y \\ -H_y & E_z & 0 & -E_x \\ -H_z & -E_y & E_x & 0 \end{pmatrix}. \tag{5.10b}$$

To obtain the Maxwell equations in the usual notation, we have merely to make the following identifications:

$$A^{\mu} = (A^0, A^m) = (\phi, \mathbf{A}), \tag{5.11a}$$

$$A_{\mu} = (A_0, A_m) = (\phi, -\mathbf{A}), \tag{5.11b}$$

and

$$j^{\mu} = (j^0, j^m) = (c\rho, \mathbf{j}), \tag{5.12a}$$

$$j_{\mu} = (j_0, j_m) = (c\rho, -\mathbf{j}). \tag{5.12b}$$

In the above equations ϕ is the scalar potential, $\mathbf{A}$ is the vector potential, ρ is the charge density, and $\mathbf{j}$ is the vector current density.

A straightforward calculation, using Eq. (5.2), gives

$$\mathbf{E} = -\nabla\phi - \frac{1}{c}\frac{\partial \mathbf{A}}{\partial t}, \tag{5.13}$$

$$\mathbf{H} = \nabla \times \mathbf{A}. \tag{5.14}$$

Equations (5.3) and (5.4), on the other hand, give

$$\nabla \cdot \mathbf{E} = 4\pi\rho, \tag{5.15}$$

$$\nabla \times \mathbf{E} = -\frac{1}{c}\frac{\partial \mathbf{H}}{\partial t}, \tag{5.16}$$

$$\nabla \cdot \mathbf{H} = 0, \tag{5.17}$$

$$\nabla \times \mathbf{H} = \frac{1}{c}\frac{\partial \mathbf{E}}{\partial t} + \frac{4\pi}{c}\mathbf{j}. \tag{5.18}$$

5.1.1 Maxwell's Equations in Curved Spacetime

It is easily seen that generalizations of these equations to curved spacetime are achieved by the following equations:

$$\nabla_\nu f^{\mu\nu} = \frac{4\pi}{c} j^\mu, \quad (5.19)$$

$$f_{\mu\nu} = \nabla_\nu A_\mu - \nabla_\mu A_\nu, \quad (5.20)$$

for Eqs. (5.3) and (5.2), whereas

$$\nabla_\delta \left[\epsilon^{\alpha\beta\gamma\delta} (-g)^{-1/2} f_{\alpha\beta} \right] = 0, \quad (5.21)$$

generalizes Eq. (5.4).

It will be noted that Eqs. (5.20) and (5.21) are identical to Eqs. (5.2) and (5.4), respectively. In Eq. (5.21) $\epsilon^{\alpha\beta\gamma\delta}$ is the totally skew-symmetric tensor density of weight $+1$ with values $+1$ and -1, depending on whether $\alpha\beta\gamma\delta$ is an even or an odd permutation of 0123 and zero otherwise. In the above equations the symbol ∇_α means covariant differentiation (see Chapter 6).

The Lagrangian density (5.1) can be extended, in the presence of gravitation, as follows:

$$\pounds = -\frac{1}{16\pi} \sqrt{-g} f_{\alpha\beta} f^{\alpha\beta} - \frac{1}{c} \sqrt{-g} j^\alpha A_\alpha + \pounds_e, \quad (5.22)$$

where now we use the curved spacetime metric tensor to raise the indices of the Maxwell tensor,

$$f_{\alpha\beta} f^{\alpha\beta} = g^{\alpha\mu} g^{\beta\nu} f_{\alpha\beta} f_{\mu\nu}. \quad (5.23)$$

The Lagrangian density (5.22) then leads to the field equation

$$\frac{\partial}{\partial x^\beta} \left[\frac{\partial \pounds}{\partial (\partial A_\alpha / \partial x^\beta)} \right] - \frac{\partial \pounds}{\partial A_\alpha} = 0. \quad (5.24)$$

Using the Lagrangian density (5.22) in the field equation (5.24) gives

$$\frac{\partial \pounds}{\partial A_\alpha} = -\frac{1}{c} \sqrt{-g} j^\alpha, \quad (5.25)$$

$$\frac{\partial \pounds}{\partial (\partial A_\alpha / \partial x^\beta)} = -\frac{1}{4\pi} \sqrt{-g} g^{\alpha\mu} g^{\beta\nu} f_{\mu\nu} = -\frac{1}{4\pi} \sqrt{-g} f^{\alpha\beta}. \quad (5.26)$$

5.1. MAXWELL'S THEORY

Accordingly, using Eq. (5.24) we obtain the following equation:

$$\frac{1}{\sqrt{-g}}\frac{\partial\left(\sqrt{-g}f^{\alpha\beta}\right)}{\partial x^\beta} = \frac{4\pi}{c}j^\alpha, \tag{5.27}$$

or

$$\nabla_\beta f^{\alpha\beta} = \frac{4\pi}{c}j^\alpha. \tag{5.28}$$

Equations (5.27) and (5.28) are Maxwell equations in the *presence* of gravitation and are a generalization of their flat-space counterpart, Eq. (5.3).

It remains to generalize the Maxwell equations (5.4) into curved spacetime. This may be achieved by replacing the partial derivatives appearing in those equations by covariant derivatives. The result is

$$\nabla_\gamma f_{\alpha\beta} + \nabla_\alpha f_{\beta\gamma} + \nabla_\beta f_{\gamma\alpha} = 0, \tag{5.29}$$

or, in its alternative form,

$$\nabla_\beta {}^*f^{\alpha\beta} = 0. \tag{5.30}$$

One can easily show that the above equations are identical to Eqs. (5.4) for the flat-space case.

Continuity Equation

The equation of continuity is obtained in the electromagnetic theory from the Maxwell equations (5.3). We obtain, because of the antisymmetry of the Maxwell field $f^{\mu\nu}$, the following:

$$\frac{\partial j^\mu}{\partial x^\mu} = 0. \tag{5.31}$$

In the presence of gravitation, on the other hand, the Maxwell equations (5.27) yield the following equation of continuity in curved spacetime:

$$\frac{\partial\left(\sqrt{-g}j^\alpha\right)}{\partial x^\alpha} = 0. \tag{5.32}$$

Again, it can easily be shown that the latter formula may also be written in the equivalent form

$$\nabla_\alpha j^\alpha = 0. \tag{5.33}$$

5.2 Spinors in Curved Spacetime

Two-component spinors are introduced in curved spacetime at each point in a "tangent" two-dimensional complex space. We then associate with every tensor a spinor. The opposite is not correct; not all spinors correspond to tensors. This is a consequence of the fact that tensors are associated with the Lorentz group, whereas spinors are associated with its covering group SL(2,C), and the correspondence between the two groups is a homomorphism rather than an isomorphism. Another way of looking at this is that the group SL(2,C) yields all representations, including those with the half-integral spins, whereas the Lorentz group yields only the representations with integral spins.

5.2.1 Correspondence between Spinors and Tensors

The correspondence between spinors and tensors is achieved by means of mixed quantities that were first introduced by Infeld and van der Waerden. These are four 2×2 Hermitian matrices, denoted by $\sigma^\mu_{AB'}$. Here Greek-letter indices are the usual spacetime indices of tensors taking the values 0, 1, 2, 3, whereas Roman capital indices are the spinor indices taking the values 0, 1. The primed indices refer to the complex conjugate and take the values $0'$, $1'$.

The hermiticity of the matrices $\sigma^\mu_{AB'}$ means, using spinor notation, that

$$\sigma^\mu_{AB'} = \overline{\sigma^\mu_{BA'}} = \overline{\sigma}^\mu_{B'A}. \tag{5.34}$$

The matrices $\sigma^\mu_{AB'}$ are generalizations of the unit matrix and the three Pauli matrices. They are functions of spacetime. There is no need to calculate them explicitly, however, when spinors are used in general relativity theory.

The relationship between the matrices σ^μ and the geometrical metric tensor $g_{\mu\nu}$ is as follows:

$$g_{\mu\nu}\sigma^\mu_{AB'}\sigma^\nu_{CD'} = \epsilon_{AC}\epsilon_{B'D'}. \tag{5.35}$$

Here ϵ_{AC} and $\epsilon_{B'D'}$, along with ϵ^{AC} and $\epsilon^{B'D'}$ to be used in the sequel, are the skew-symmetric Levi-Civita metric spinors. They are defined by the matrix

$$\epsilon = \begin{pmatrix} 0 & 1 \\ -1 & 0 \end{pmatrix}. \tag{5.36}$$

The raising and lowering of spinor indices is accomplished by means of the above metric spinors as follows.

5.2. SPINORS IN CURVED SPACETIME

5.2.2 Raising and Lowering Spinor Indices

The role of the Levi-Civita metric spinors in raising and lowering spinor indices is similar to that of the geometrical metric tensors in raising and lowering spacetime indices. There is a difference, however, since the metric spinors are antisymmetric. We will use the convention according to which

$$\xi^A = \epsilon^{AB}\xi_B, \quad \eta^{A'} = \epsilon^{A'B'}\eta_{B'}, \qquad (5.37)$$

and

$$\xi_A = \xi^B \epsilon_{BA}, \quad \eta_{A'} = \eta^{B'}\epsilon_{B'A'}, \qquad (5.38)$$

for arbitrary spinors ξ and η. The above formulas give, for instance, $\xi^0 = \xi_1$ and $\xi^1 = -\xi_0$. We have, moreover,

$$\xi^A \eta_A = \xi^A \eta^B \epsilon_{BA} = -\xi^A \epsilon_{AB}\eta^B = -\xi_B \eta^B. \qquad (5.39)$$

Hence contraction with spinor indices should be done according to the convention given by Eqs. (5.37) and (5.38).

We finally notice that the Levi-Civita spinor satisfies the simple identity

$$\epsilon_{AB}\epsilon_{CD} + \epsilon_{AC}\epsilon_{DB} + \epsilon_{AD}\epsilon_{BC} = 0. \qquad (5.40)$$

The above identity may easily be proved by taking the different values of the indices A, B, C, and D.

5.2.3 Properties of the σ Matrices

In addition to relation (5.35), which the Hermitian matrices σ^μ satisfy, they also satisfy the following formulas:

$$\sigma^\mu_{AB'}\sigma^{\nu AB'} = g^{\mu\nu}, \qquad (5.41a)$$

or

$$\sigma^\mu_{AB'}\sigma^{AB'}_\nu = \delta^\mu_\nu, \qquad (5.41b)$$

which is equivalent to Eq. (5.41a).

The spinor equivalent of a tensor is a quantity that has an unprimed and a primed spinor index for each spacetime tensor index. The spinor equivalent of the tensor $T_{\alpha\beta}$, for instance, is given by

$$T_{AB'CD'} = \sigma^\alpha_{AB'}\sigma^\beta_{CD'}T_{\alpha\beta}. \qquad (5.42)$$

The above formula may be reversed. We then obtain the tensor corresponding to the spinor $T_{AB'CD'}$. We obtain

$$\sigma_\mu^{AB'}\sigma_\nu^{CD'}T_{AB'CD'} = \sigma_\mu^{AB'}\sigma_\nu^{CD'}\sigma_{AB'}^\alpha\sigma_{CD'}^\beta T_{\alpha\beta}$$
$$= \delta_\mu^\alpha\delta_\nu^\beta T_{\alpha\beta} = T_{\mu\nu}, \tag{5.43}$$

by Eqs. (5.41).

The spinor equivalent of the geometrical metric tensor $g_{\mu\nu}$ is given by

$$g_{AB'CD'} = \sigma_{AB'}^\alpha\sigma_{CD'}^\beta g_{\alpha\beta} = \epsilon_{AC}\epsilon_{B'D'}, \tag{5.44a}$$

$$g^{AB'CD'} = \sigma_\alpha^{AB'}\sigma_\beta^{CD'} g^{\alpha\beta} = \epsilon^{AC}\epsilon^{B'D'}, \tag{5.44b}$$

by Eq. (5.35). The above spinors are, in fact, the usual flat spacetime metric tensors, but are now having the form

$$g^{AB'CD'} = g_{AB'CD'} = \begin{pmatrix} 0 & & & 0 & 1 \\ & & & -1 & 0 \\ 0 & -1 & & & 0 \\ 1 & 0 & & & \end{pmatrix}, \tag{5.45a}$$

rather than that of the Minkowskian metric tensor $\eta_{\alpha\beta}$.

The indices of the matrix (5.45a) are arranged in such a way that the first pair, $AB' = 00', 01', 10', 11'$, denotes the rows, whereas the second pair CD', taking the same values, denotes the columns. We also have

$$g_{AB'}^{CD'} = \sigma_{\alpha AB'}\sigma^{\alpha CD'} = \delta_{AB'}^{CD'} = \delta_A^C\delta_{B'}^{D'} = \begin{pmatrix} 1 & 0 & & 0 \\ 0 & 1 & & \\ & & 1 & 0 \\ 0 & & 0 & 1 \end{pmatrix}, \tag{5.45b}$$

which is equivalent to Eq. (5.45a).

5.2.4 The Metric $g_{AB'CD'}$ and the Minkowskian Metric $\eta_{\mu\nu}$

The relationship between the flat spacetime metric $g_{AB'CD'}$ and the ordinary Minkowskian metric tensor $\eta_{\mu\nu}$ is obtained if we take for the matrices $\sigma_{AB'}^\mu$ the Pauli matrices and the unit matrix, divided by $\sqrt{2}$. Accordingly we may take

$$\sigma_{AB'}^0 = \frac{1}{\sqrt{2}}\begin{pmatrix} 1 & 0 \\ 0 & 1 \end{pmatrix}, \quad \sigma_{AB'}^1 = \frac{1}{\sqrt{2}}\begin{pmatrix} 0 & 1 \\ 1 & 0 \end{pmatrix},$$

5.2. SPINORS IN CURVED SPACETIME

$$\sigma^2_{AB'} = \frac{1}{\sqrt{2}} \begin{pmatrix} 0 & i \\ -i & 0 \end{pmatrix}, \quad \sigma^3_{AB'} = \frac{1}{\sqrt{2}} \begin{pmatrix} 1 & 0 \\ 0 & -1 \end{pmatrix}, \quad (5.46a)$$

and, raising the indices AB', we obtain

$$\sigma^{0AB'} = \frac{1}{\sqrt{2}} \begin{pmatrix} 1 & 0 \\ 0 & 1 \end{pmatrix}, \quad \sigma^{1AB'} = \frac{1}{\sqrt{2}} \begin{pmatrix} 0 & -1 \\ -1 & 0 \end{pmatrix},$$

$$\sigma^{2AB'} = \frac{1}{\sqrt{2}} \begin{pmatrix} 0 & i \\ -i & 0 \end{pmatrix}, \quad \sigma^{3AB'} = \frac{1}{\sqrt{2}} \begin{pmatrix} -1 & 0 \\ 0 & 1 \end{pmatrix}. \quad (5.46b)$$

The tensor equivalent to the spinor $g_{AB'CD'}$ is then given by

$$\eta_{\mu\nu} = \sigma^{AB'}_\mu \sigma^{CD'}_\nu g_{AB'CD'} = \begin{pmatrix} +1 & & & \\ & -1 & & \\ & & -1 & \\ & & & -1 \end{pmatrix}, \quad (5.47)$$

which is, of course, the Minkowskian metric tensor.

5.2.5 Hermitian Spinors

When taking complex conjugate of a spinor, unprimed indices become primed, and vice versa. The complex conjugate of the spinor $S_{AB'}$, for instance, is given by

$$\overline{S_{AB'}} = \overline{S}_{A'B}. \quad (5.48)$$

When a tensor is real, its spinor equivalent is Hermitian. Suppose, for instance, that V_α is a real vector and its spinor equivalent is $V_{AB'}$. Then we have

$$\overline{V}_{B'A} = \overline{V_{BA'}} = \overline{\sigma^\alpha_{BA'} V_\alpha} = \overline{\sigma}^\alpha_{B'A} V_\alpha = \sigma^\alpha_{AB'} V_\alpha = V_{AB'}, \quad (5.49)$$

by Eq. (5.34).

If the vector V_α is null, namely, $V_\alpha V^\alpha = g_{\alpha\beta} V^\alpha V^\beta = 0$, then its corresponding spinor can be given as a product of an unprimed spinor with a primed spinor,

$$V_{AB'} = \alpha_A \beta_B. \quad (5.50)$$

If, moreover, the vector V_α is real, then $\beta_{B'}$ is a multiple of $\overline{\alpha}_{B'}$,

$$V_{AB'} = \alpha_A \overline{\alpha}_{B'}. \quad (5.51)$$

Any direction along the light cone, therefore, corresponds uniquely to a one-index spinor *ray*, namely, to a set of spinors proportional to a given spinor.

5.3 Covariant Derivative of a Spinor

The covariant derivative of a spinor ξ_A, denoted by $\nabla_\mu \xi_A$, is given by

$$\nabla_\mu \xi_A = \frac{\partial \xi_A}{\partial x^\mu} - \Gamma^B_{A\mu} \xi_B. \tag{5.52}$$

Here $\Gamma^B_{A\mu}$ is the *spinor affine connection*. When taking the covariant derivative of the complex conjugate of the spinor ξ_A, we have

$$\nabla_\mu \bar{\xi}_{A'} = \frac{\partial \bar{\xi}_{A'}}{\partial x^\mu} - \bar{\Gamma}^{B'}_{A'\mu} \bar{\xi}_{B'}. \tag{5.53}$$

Analogous equations hold for the spinors ξ^A and $\bar{\xi}^{A'}$,

$$\nabla_\mu \xi^A = \frac{\partial \xi^A}{\partial x^\mu} + \Gamma^A_{B\mu} \xi^B. \tag{5.54}$$

$$\nabla_\mu \bar{\xi}^{A'} = \frac{\partial \bar{\xi}^{A'}}{\partial x^\mu} + \bar{\Gamma}^{A'}_{B'\mu} \bar{\xi}^{B'}. \tag{5.55}$$

Generalizations of the above formulas to spinors with more than one index are done similarly as for tensors. Thus we have for a spinor with two indices,

$$\nabla_\mu \eta^{AB'} = \frac{\partial \eta^{AB'}}{\partial x^\mu} + \Gamma^A_{C\mu} \eta^{CB'} + \bar{\Gamma}^{B'}_{C'\mu} \bar{\eta}^{AC'}, \tag{5.56}$$

for instance.

5.3.1 Spinor Affine Connections

The spinor affine connections introduced above are fixed by the requirement that the covariant derivatives of the matrices σ^μ and the Levi-Civita spinors all vanish,

$$\nabla_\alpha \sigma^\mu_{AB'} = 0, \tag{5.57}$$

$$\nabla_\alpha \epsilon_{AB} = 0, \quad \nabla_\alpha \epsilon^{AB} = 0, \tag{5.58a}$$

$$\nabla_\alpha \epsilon_{A'B'} = 0, \quad \nabla_\alpha \epsilon^{A'B'} = 0. \tag{5.58b}$$

The vanishing of the covariant derivative of ϵ_{AB}, for instance, implies

$$\nabla_\alpha \epsilon_{AB} = \frac{\partial \epsilon_{AB}}{\partial x^\alpha} - \Gamma^C_{A\alpha} \epsilon_{CB} - \Gamma^C_{B\alpha} \epsilon_{AC} = 0. \tag{5.59}$$

5.3. COVARIANT DERIVATIVE OF A SPINOR

Hence we obtain

$$\Gamma^C_{A\alpha}\epsilon_{CB} = \Gamma^C_{B\alpha}\epsilon_{CA}, \qquad (5.60a)$$

or

$$\Gamma_{BA\alpha} = \Gamma_{AB\alpha}, \qquad (5.60b)$$

where $\Gamma_{AB\alpha} = \Gamma^C_{B\alpha}\epsilon_{CA}$.

In the sequel we use the covariant derivative operator $\nabla_{AB'}$ defined by

$$\nabla_{AB'} = \sigma^\mu_{AB'}\nabla_\mu = \begin{pmatrix} \nabla_{00'} & \nabla_{01'} \\ \nabla_{10'} & \nabla_{11'} \end{pmatrix}. \qquad (5.61)$$

Of course the two components $\nabla_{00'}$ and $\nabla_{11'}$ are real, whereas $\nabla_{01'}$ and $\nabla_{10'}$ are complex, each being the complex conjugate to the other one, $\nabla_{10'} = \overline{\nabla_{0'1}} = \overline{\nabla_{01'}}$. The operator $\nabla_{AB'}$ is often denoted as follows:

$$\nabla_{AB'} = \begin{pmatrix} D & \delta \\ \bar{\delta} & \Delta \end{pmatrix}. \qquad (5.62)$$

In flat spacetime, and when the matrices σ^μ are presented as in Eqs. (5.46), the operator $\nabla_{AB'}$ has the simple presentation

$$\nabla_{AB'} = \frac{1}{\sqrt{2}}\begin{pmatrix} \partial_t + \partial_z & \partial_x + i\partial_y \\ \partial_x - i\partial_y & \partial_t - \partial_z \end{pmatrix}. \qquad (5.63)$$

In the above formula it has been assumed that the coordinates are Cartesian, $x^0 = t$, $x^1 = x$, $x^2 = y$, and $x^3 = z$, with the speed of light $c = 1$.

5.3.2 Spin Covariant Derivative

As was shown in Chapter 4, two-component spinors are obtained as (complex) numbers in the representation formula of the group SL(2,C). When applied in physics they become functions of spacetime and subsequently one needs to define the (coordinate) covariant derivative as was done previously. Such derivatives have a vectorial form ∇_μ or a spin form $\nabla_{AB'}$. The question arises as to whether one can define spin covariant derivatives of the form ∇_A and $\nabla_{A'}$ which are more basic than the previous ones. In the following such a derivative is presented.

To this end one considers two-component spinors as functions of some complex variables. More definitely it will be assumed that they are functions of two complex variables, z^0 and z^1, and their complex conjugates. This is equivalent to four real variables which might or might not have any relationship to spacetime coordinates. The two complex variables will be denoted collectively by z^A with $A = 0, 1$.

The spin covariant derivative may then be defined by

$$\nabla_A \zeta_B = \frac{\partial \zeta_B}{\partial z^A} - \Gamma^C_{AB} \zeta_C, \qquad (5.64a)$$

$$\nabla_A \bar{\zeta}_{B'} = \frac{\partial \bar{\zeta}_{B'}}{\partial z^A} - \bar{\Gamma}^{C'}_{AB'} \bar{\zeta}_{C'}, \qquad (5.64b)$$

$$\bar{\nabla}_{A'} \zeta_B = \frac{\partial \zeta_B}{\partial \bar{z}^{A'}} - \Gamma^C_{A'B} \zeta_C, \qquad (5.64c)$$

$$\bar{\nabla}_{A'} \bar{\zeta}_{B'} = \frac{\partial \bar{\zeta}_{B'}}{\partial \bar{z}^{A'}} - \bar{\Gamma}^{C'}_{A'B'} \bar{\zeta}_{C'}, \qquad (5.64d)$$

for the spinors ζ_B and $\bar{\zeta}_{B'}$. Similar formulas for the spin covariant derivatives of the spinors ζ^B and $\bar{\zeta}^{B'}$:

$$\nabla_A \zeta^B = \frac{\partial \zeta^B}{\partial z^A} + \Gamma^B_{AC} \zeta^C, \qquad (5.65a)$$

$$\nabla_A \bar{\zeta}^{B'} = \frac{\partial \bar{\zeta}^{B'}}{\partial z^A} + \bar{\Gamma}^{B'}_{AC'} \bar{\zeta}^{C'}, \qquad (5.65b)$$

$$\bar{\nabla}_{A'} \zeta^B = \frac{\partial \zeta^B}{\partial \bar{z}^{A'}} + \Gamma^B_{A'C} \zeta^C, \qquad (5.65c)$$

$$\bar{\nabla}_{A'} \bar{\zeta}^{B'} = \frac{\partial \bar{\zeta}^{B'}}{\partial \bar{z}^{A'}} + \bar{\Gamma}^{B'}_{A'C'} \bar{\zeta}^{C'}, \qquad (5.65d)$$

It will be noted that Eqs. (5.64c) and (5.65c) are the complex conjugates of Eqs. (5.64b) and (5.65b), respectively. The quantities Γ are called *spin affine connections*.

We finally relate the spin covariant derivative to the ordinary coordinate covariant derivatives. This is done by expressing the ordinary covariant derivative as a product of the newly defined ones,

$$\nabla_{AB'} = \nabla_A \bar{\nabla}_{B'}, \qquad (5.66)$$

where $\nabla_{AB'} = \sigma^\mu_{AB'} \nabla_\mu$, and thus one has

$$\nabla_\mu \sigma^{AB'}_\mu \nabla_{AB'} = \sigma^{AB'}_\mu \nabla_A \bar{\nabla}_{B'}. \qquad (5.67)$$

Since $\nabla_{AB'}$ is Hermitian, ∇_μ is real. In terms of the complex variables z^0 and z^1, one then has

$$\nabla_{AB'} = \begin{pmatrix} \nabla_{z^0} \nabla_{\bar{z}^0} & \nabla_{z^0} \nabla_{\bar{z}^1} \\ \nabla_{z^1} \nabla_{\bar{z}^0} & \nabla_{z^1} \nabla_{\bar{z}^1} \end{pmatrix}. \qquad (5.68)$$

5.3.3 A Useful Formula

Finally it is worthwhile mentioning that any spinor with two indices ξ_{AB} satisfies the relation

$$2\xi_{[AB]} = \xi_{AB} - \xi_{BA} = \xi_C^C \epsilon_{AB}, \tag{5.69}$$

where $\xi_C^C = \epsilon^{CD}\xi_{CD}$. Equation (5.69) is a consequence of the identity (5.40) and is obtained from it by multiplying it by ξ^{CD}. Formulas similar to Eq. (5.69) are valid for spinors having more than two indices. Thus we have, for instance,

$$2S_{[AB]CD} = S_{ABCD} - S_{BACD} = \epsilon_{AB}S_F{}^F{}_{CD}, \tag{5.70}$$

for an arbitrary spinor S_{ABCD}.

In the next section we apply the theory presented above to the electromagnetic field.

5.4 The Electromagnetic Field Spinors

We may now apply the theory presented in the last section to the electromagnetic and gravitational fields. Accordingly, all the field variables will be presented in spinorial form. This includes the electromagnetic field potential and tensor, the gravitational field curvature tensor, the Weyl conformal tensor, and the Ricci and Einstein tensors. This section is devoted to the electromagnetic field.

5.4.1 The Electromagnetic Potential Spinor

The spinor equivalent to the electromagnetic potential A_μ is the spinor $A_{CD'}$ given by

$$A_{CD'} = \sigma^\mu_{CD'} A_\mu. \tag{5.71}$$

Since the vector A_μ is real, the spinor $A_{CD'}$ is Hermitian, namely, $A_{CD'} = \overline{A_{D'C}}$. Thus the components $A_{00'}$ and $A_{11'}$ are real, whereas $A_{01'}$ and $A_{10'}$ are complex conjugate to each other,

$$A_{10'} = \overline{A_{01'}} = \overline{A}_{0'1}. \tag{5.72}$$

5.4.2 The Electromagnetic Field Spinor

The spinor equivalent to the electromagnetic field tensor $f_{\mu\nu}$ is given by

$$f_{AB'CD'} = \sigma^{\mu}_{AB'}\sigma^{\nu}_{CD'}f_{\mu\nu}. \tag{5.73}$$

Since $f_{\mu\nu}$ is skew-symmetric and real, the spinor (5.73) is antisymmetric under the exchange of the two pairs of indices AB' and CD' and is Hermitian with respect to these indices. Accordingly we have

$$f_{AB'CD'} = -f_{CD'AB'}, \tag{5.74}$$

$$f_{AB'CD'} = \overline{f}_{B'AD'C}. \tag{5.75}$$

Because of the antisymmetry property of the electromagnetic field, we may decompose its spinor equivalent (5.73). To this end we present Eq. (5.73) as follows:

$$f_{AB'CD'} = \frac{1}{2}\left(f_{AB'CD'} - f_{CB'AD'}\right) + \frac{1}{2}\left(f_{CB'AD'} - f_{CD'AB'}\right). \tag{5.76}$$

Here the first expression in parentheses is skew-symmetric in the indices A and C, while the second expression is skew-symmetric in the indices B' and D'. According to Eq. (5.70) we then have

$$f_{AB'CD'} - f_{CB'AD'} = \epsilon_{AC} f_{GB'}{}^{G}{}_{D'}, \tag{5.77}$$

$$f_{CB'AD'} - f_{CD'AB'} = \epsilon_{B'D'} f_{CG'A}{}^{G'}. \tag{5.78}$$

Hence Eq. (5.76) may now be written in the form

$$f_{AB'CD'} = \frac{1}{2}\left(\epsilon_{AC} f_{GB'}{}^{G}{}_{D'} + \epsilon_{B'D'} f_{CG'A}{}^{G'}\right). \tag{5.79}$$

5.4.3 Decomposition of the Electromagnetic Spinor

Let us now denote

$$\phi_{AB} = \frac{1}{2} f_{AC'B}{}^{C'}. \tag{5.80}$$

We then have

$$\phi_{BA} = \frac{1}{2} f_{BC'A}{}^{C'} = -\frac{1}{2} f_A{}^{C'}{}_{BC'} = \frac{1}{2} f_{AC'B}{}^{C'} = \phi_{AB}. \tag{5.81}$$

5.4. THE ELECTROMAGNETIC FIELD SPINORS

Hence ϕ_{AB} is a symmetric spinor. Taking now the complex conjugate of ϕ_{AB} we obtain

$$\overline{\phi}_{A'B'} = \overline{\phi_{AB}} = \frac{1}{2}\overline{f_{AC'B}}^{C'} = \frac{1}{2}\overline{f}_{A'CB'}^{\ \ C} = \frac{1}{2}f_{CA'}^{\ \ C}{}_{B'}, \qquad (5.82)$$

where use has been made of the hermiticity property of the spinor $f_{AB'CD'}$. As a result, Eq. (5.79) may now be written in the form

$$f_{AB'CD'} = \epsilon_{AC}\overline{\phi}_{B'D'} + \phi_{AC}\epsilon_{B'D'}, \qquad (5.83)$$

where use has been made of Eqs. (5.81) and (5.82).

We thus see that the electromagnetic field tensor $f_{\mu\nu}$ is equivalent to the symmetric spinor ϕ_{AB}. The six real components of $f_{\mu\nu}$ are presented by the three complex components of ϕ_{AB}. These are ϕ_{00}, $\phi_{01} = \phi_{10}$, and ϕ_{11}. The spinor ϕ_{AB} will be referred to by us as the *electromagnetic field spinor*, or simply the *Maxwell spinor*. In the sequel use will be made of the notation

$$\phi_0 = \phi_{00}, \quad \phi_1 = \phi_{01} = \phi_{10}, \quad \phi_2 = \phi_{11}. \qquad (5.84)$$

We finally find the spinor equivalent to the dual to the electromagnetic field tensor. If $f_{\mu\nu}$ is the electromagnetic field tensor and $*f_{\alpha\beta}$ is its dual,

$$*f_{\alpha\beta} = \frac{1}{2}\sqrt{-g}\epsilon_{\alpha\beta\gamma\delta}f^{\gamma\delta}, \qquad (5.85)$$

where $\epsilon_{\alpha\beta\gamma\delta}$ is the Levi-Civita tensor density of weight $W = -1$, then the spinor equivalent to the dual is given by

$$*f_{AB'CD'} = i\left(\epsilon_{AC}\overline{\phi}_{B'D'} - \phi_{AC}\epsilon_{B'D'}\right). \qquad (5.86)$$

5.4.4 Intrinsic Spin Structure

We now apply the commutator $(\nabla_N\nabla_M - \nabla_M\nabla_N)$ to an arbitrary spinor ξ_Q. A simple calculation then yields

$$(\nabla_N\nabla_M - \nabla_M\nabla_N)\xi_Q = -\Phi^P{}_{QMN}\xi_P, \qquad (5.87)$$

where

$$\Phi^P{}_{QMN} = \Gamma^P_{QM,N} - \Gamma^P_{QN,M} + \Gamma^B_{QM}\Gamma^P_{BN} - \Gamma^B_{QN}\Gamma^P_{BM}. \qquad (5.88)$$

In Eq. (5.88) a comma followed by a capital letter means a partial derivative, $f_{,M} = \partial f/\partial z^M$ (see Subsection 5.3.2). In the same way we obtain

$$(\nabla_N\nabla_M - \nabla_M\nabla_N)\xi^Q = \Phi^Q{}_{PMN}\xi^P. \qquad (5.89)$$

The above formulas are analogous to those for defining the Riemann curvature tensor (see next chapter).

From Eq. (5.87) we obtain

$$(\nabla_N \nabla_M - \nabla_M \nabla_N)\xi_Q = \Phi_{PQMN}\xi^P. \tag{5.90}$$

On the other hand, by lowering the free index Q in Eq. (5.89), we have

$$(\nabla_N \nabla_M - \nabla_M \nabla_N)\xi_Q = \Phi_{QPMN}\xi^P. \tag{5.91}$$

Comparing the last two formulas we find that

$$\Phi_{PQMN} = \Phi_{QPMN}, \tag{5.92}$$

thus the spinor Φ is symmetric with its first two indices. It is also obvious, by its derivation, that Φ is skew-symmetric with respect to its last two indices,

$$\Phi_{PQMN} = -\Phi_{PQNM}. \tag{5.93}$$

Using now Eq. (5.70), one thus has

$$\Phi_{PQMN} = \Phi_{PQ}\,\epsilon_{MN}, \tag{5.94}$$

where $\Phi_{PQ} = \Phi_{QP} = \Phi_{PQA}{}^A$.

The spinor Φ_{PQ} thus has the same symmetry properties as the electromagnetic field spinor ϕ_{PQ} (see Subsection 5.4.3).

5.4.5 Pauli, Dirac and Maxwell Equations

1. The neutrino equation

The neutrino is described by a two-component Pauli spinor, η^A. The neutrino field equation is then given by

$$\nabla_{AB'}\eta^A = 0. \tag{5.95}$$

2. The Dirac equation

The Dirac equation for a spin-$\frac{1}{2}$ particle is given by

$$\gamma^\mu\left(\nabla_\mu - ieA_\mu\right)\psi = m\psi,$$

in units in which $\hbar = c = 1$. Here ψ is a four-component spinor and γ^μ are four 4×4 matrices that satisfy certain commutation relations. The

5.4. THE ELECTROMAGNETIC FIELD SPINORS

4-component spinor ψ is composed of two 2-component spinors α^A and β^A as follows:

$$\psi = \begin{pmatrix} \alpha_0 \\ \alpha_1 \\ \overline{\beta}^0 \\ \overline{\beta}^1 \end{pmatrix}.$$

The Dirac equation can also be described by coupled two equations with the two 2-component spinors α^A and β^A (see Problem 5.12):

$$(\nabla_{CD'} + ieA_{CD'})\alpha^C = \frac{m}{\sqrt{2}}\overline{\beta}_{D'}, \tag{5.96a}$$

$$(\nabla_{CD'} + ieA_{CD'})\beta^C = \frac{m}{\sqrt{2}}\overline{\alpha}_{D'}, \tag{5.96b}$$

where $A_{CD'} = \sigma^\mu_{CD'} A_\mu$, with A_μ being the electromagnetic potential, and m and e are the mass and the charge of the particle.

3. The Maxwell equations

These are given by

$$\nabla_\nu f^{\mu\nu} = \frac{4\pi}{c} j^\mu, \tag{5.97a}$$

$$\nabla_\nu {}^* f^{\mu\nu} = 0, \tag{5.97b}$$

where ${}^*f$ is the dual to f, and c is the speed of light in vacuum.

The spinor version of the above equations is

$$\nabla_{CD'} f^{AB'CD'} = \frac{4\pi}{c} j^{AB'}, \tag{5.98a}$$

$$\nabla_{CD'} {}^* f^{AB'CD'} = 0, \tag{5.98b}$$

where f and ${}^*f$ are given by

$$f^{AB'CD'} = \epsilon^{AC}\overline{\phi}^{B'D'} + \phi^{AC}\epsilon^{B'D'}, \tag{5.99a}$$

$${}^* f^{AB'CD'} = i\left(\epsilon^{AC}\overline{\phi}^{B'D'} - \phi^{AC}\epsilon^{B'D'}\right). \tag{5.99b}$$

Defining now the spinor $f^+ = f + i\,{}^*f$, we then obtain for the Maxwell equations

$$\nabla_{CD'} f^{+AB'CD'} = \frac{4\pi}{c} j^{AB'}, \tag{5.100}$$

with $f^+_{AB'CD'} = 2\phi_{AC}\epsilon_{B'D'}$.

In the next chapter we discuss the gravitational field dynamical variables, starting with a review of general relativity theory.

5.5 Problems

5.1 Decompose the commutator of the covariant differentiation operators $(\nabla_\nu \nabla_\mu - \nabla_\mu \nabla_\nu)$ in spinor form.

Solution: The spinor equivalent to the commutator $(\nabla_\nu \nabla_\mu - \nabla_\mu \nabla_\nu)$ is given by
$$(\nabla_{CD'}\nabla_{AB'} - \nabla_{AB'}\nabla_{CD'}). \tag{1}$$

By adding and subtracting identical terms we obtain

$$\nabla_{CD'}\nabla_{AB'} - \nabla_{AB'}\nabla_{CD'}$$
$$= (\nabla_{CD'}\nabla_{AB'} - \nabla_{CB'}\nabla_{AD'}) + (\nabla_{CB'}\nabla_{AD'} - \nabla_{AB'}\nabla_{CD'}). \tag{2}$$

The first bracket on the right-hand side is antisymmetric in the indices D' and B', whereas the second bracket is antisymmetric in the indices C and A. Hence we can use Eq. (5.69). We then obtain

$$\nabla_{CD'}\nabla_{AB'} - \nabla_{AB'}\nabla_{CD'} = \epsilon_{D'B'}\nabla_{(AC)} + \epsilon_{CA}\nabla_{(B'D')}, \tag{3}$$

where use has been made of the notation

$$\nabla_{(AC)} = \frac{1}{2}\left(\nabla_{AE'}\nabla_C^{E'} + \nabla_{CE'}\nabla_A^{E'}\right), \tag{4}$$

$$\nabla_{(B'D')} = \frac{1}{2}\left(\nabla_{EB'}\nabla^E_{D'} + \nabla_{ED'}\nabla^E_{B'}\right). \tag{5}$$

5.2 Show that the Levi-Civita contravariant tensor density $\epsilon^{\mu\nu\alpha\beta}$ of weight $W = +1$ may be presented in the form

$$\epsilon^{\mu\nu\alpha\beta} = i\sqrt{-g}\sigma^{\mu CB'}\sigma^{\nu AD'}\left(\sigma^\alpha_{CD'}\sigma^\beta_{AB'} - \sigma^\beta_{CD'}\sigma^\alpha_{AB'}\right). \tag{1}$$

Solution: We start with the identity

$$\epsilon^{MN'PQ'AB'CD'} = -\left(\epsilon^{AM}\epsilon^{B'Q'}\epsilon^{CP}\epsilon^{D'N'} - \epsilon^{AP}\epsilon^{B'N'}\epsilon^{CM}\epsilon^{D'Q'}\right), \tag{2}$$

where $\epsilon^{MN'PQ'AB'CD'}$ is the spinor equivalent to the Levi-Civita tensor density $\epsilon^{\mu\nu\alpha\beta}$. Identity (2) may be verified by checking the components of Eq. (2) for the various values of its indices. The indices MN', PQ', AB', CD' take the values $00'$, $01'$, $10'$, $11'$, and $\epsilon^{MN'PQ'AB'CD'}$ takes the values $+1$ and -1, depending upon whether MN', PQ', AB', CD' is an even or an odd permutation of $00'$, $01'$, $10'$, $11'$, and zero otherwise.

5.5. PROBLEMS

Using now the relation

$$\epsilon^{\mu\nu\alpha\beta}\sigma_\mu^{MN'}\sigma_\nu^{PQ'}\sigma_\alpha^{AB'}\sigma_\beta^{CD'} = \sigma\epsilon^{MN'PQ'AB'CD'}, \qquad (3)$$

where

$$\sigma = \det\sigma_\mu^{AB'} = \sqrt{g} = -i\sqrt{-g}, \qquad (4)$$

Eq. (3) then yields, using Eq. (2),

$$\epsilon^{\mu\nu\alpha\beta}\sigma_\mu^{MN'}\sigma_\nu^{PQ'}\sigma_\alpha^{AB'}\sigma_\beta^{CD'}$$
$$= i\sqrt{-g}\left(\epsilon^{AM}\epsilon^{B'Q'}\epsilon^{CP}\epsilon^{D'N'} - \epsilon^{AP}\epsilon^{B'N'}\epsilon^{CM}\epsilon^{D'Q'}\right). \qquad (5)$$

Multiplying now the latter formula by $\sigma^\rho_{MN'}\sigma^\tau_{PQ'}\sigma^\gamma_{AB'}\sigma^\delta_{CD'}$ then yields

$$\epsilon^{\rho\tau\gamma\delta} = i\sqrt{-g}\sigma^\rho_{MN'}\sigma^\tau_{PQ'}\left(\sigma^{\gamma MQ'}\sigma^{\delta PN'} - \sigma^{\delta MQ'}\sigma^{\gamma PN'}\right). \qquad (6)$$

Equation (6) is identical to Eq. (1) if we raise the indices MN' and PQ' of $\sigma^\rho_{MN'}$ and $\sigma^\tau_{PQ'}$ and lower the same indices of the bracket without causing any change.

5.3 Show that the spinor equivalent to the tensor

$$\epsilon^{\alpha\beta}_{\mu\nu} = \sqrt{-g}g^{\alpha\rho}g^{\beta\sigma}\epsilon_{\rho\sigma\mu\nu} = \frac{1}{\sqrt{-g}}g_{\mu\rho}g_{\nu\sigma}\epsilon^{\rho\sigma\alpha\beta}, \qquad (1)$$

where $\epsilon_{\rho\sigma\mu\nu}$ and $\epsilon^{\rho\sigma\alpha\beta}$ are the Levi-Civita covariant and contravariant tensor densities of weights $W = -1$ and $+1$, respectively, is given by

$$\epsilon^{AB'CD'}_{EF'GH'} = i\left(\delta^A_E\delta^C_G\delta^{B'}_{H'}\delta^{D'}_{F'} - \delta^A_G\delta^C_E\delta^{B'}_{F'}\delta^{D'}_{H'}\right). \qquad (2)$$

Solution: Using the representation for the Levi-Civita tensor density given by Eq. (1) of the previous problem, we obtain

$$\epsilon^{\alpha\beta}_{\gamma\delta} = i\sigma^{CB'}_\gamma\sigma^{AD'}_\delta\left(\sigma^\alpha_{CD'}\sigma^\beta_{AB'} - \sigma^\alpha_{AB'}\sigma^\beta_{CD'}\right). \qquad (3)$$

The spinor equivalent of Eq. (3) is

$$\epsilon^{EF'GH'}_{IJ'KL'} = i\delta^{CB'}_{IJ'}\delta^{AD'}_{KL'}\left(\delta^{EF'}_{CD'}\delta^{GH'}_{AB'} - \delta^{EF'}_{AB'}\delta^{GH'}_{CD'}\right), \qquad (4)$$

which may also be written as

$$\epsilon^{EF'GH'}_{IJ'KL'} = i\left(\delta^{EF'}_{IL'}\delta^{GH'}_{KJ'} - \delta^{EF'}_{KJ'}\delta^{GH'}_{IL'}\right). \qquad (5)$$

In the above formulas use has been made of the notation
$$\delta^{AB'}_{CD'} = \delta^A_C \delta^{B'}_{D'}. \tag{6}$$

5.4 Show that the spinors equivalent to the tensor $\epsilon^{\alpha\beta}_{\gamma\delta}$ and to the tensor $\delta^{\alpha\beta}_{\gamma\delta}$ are related by
$$\epsilon^{AB'CD'}_{EF'GH'} = i\delta^{AB'CD'}_{EH'GF'}. \tag{1}$$

Solution: The tensor $\delta^{\alpha\beta}_{\gamma\delta}$ is given by
$$\delta^{\alpha\beta}_{\mu\nu} = \delta^\alpha_\mu \delta^\beta_\nu - \delta^\alpha_\nu \delta^\beta_\mu. \tag{2}$$

Hence its spinor equivalent is given by
$$\delta^{AB'CD'}_{EF'GH'} = \delta^{AB'}_{EF'} \delta^{CD'}_{GH'} - \delta^{AB'}_{GH'} \delta^{CD'}_{EF'}, \tag{3}$$

where
$$\delta^{AB'}_{EF'} = \delta^A_E \delta^{B'}_{F'}. \tag{4}$$

Comparing Eq. (3) with Eq. (5) of the previous problem for the spinor equivalent of the tensor $\epsilon^{\alpha\beta}_{\mu\nu}$, we obtain Eq. (1).

5.5 Write the spinor affine connections in terms of the ordinary tensor affine connections and the matrices σ^α and their derivatives.

Solution: We use the fact that the covariant derivative of the matrices σ^α vanish. We then obtain
$$\nabla_\mu \sigma^\nu_{AB'} = \partial_\mu \sigma^\nu_{AB'} + \Gamma^\nu_{\mu\rho} \sigma^\rho_{AB'} - \Gamma^C_{A\mu} \sigma^\nu_{CB'} - \overline{\Gamma}^{D'}_{B'\mu} \sigma^\nu_{AD'} = 0, \tag{1}$$

where $\partial_\mu f = \partial f/\partial x^\mu$. Multiplying now the above equation by $\sigma^{EF'}_\nu$ and summing over the index ν, we obtain
$$\sigma^{EF'}_\nu \partial_\mu \sigma^\nu_{AB'} + \sigma^{EF'}_\nu \Gamma^\nu_{\mu\rho} \sigma^\rho_{AB'} - \Gamma^E_{A\mu} \delta^{F'}_{B'} - \overline{\Gamma}^{D'}_{B'\mu} \delta^E_A = 0. \tag{2}$$

Contracting now the indices F' and B' in Eq. (2), yields
$$\sigma^{EB'}_\nu \partial_\mu \sigma^\nu_{AB'} + \sigma^{EB'}_\nu \Gamma^\nu_{\mu\rho} \sigma^\rho_{AB'} - 2\Gamma^E_{A\mu} = 0, \tag{3}$$

and therefore
$$\Gamma^E_{A\mu} = \frac{1}{2} \sigma^{EB'}_\nu \left(\Gamma^\nu_{\mu\rho} \sigma^\rho_{AB'} + \partial_\mu \sigma^\nu_{AB'} \right). \tag{4}$$

Likewise we obtain, by contracting indices A and E in Eq. (2),
$$\overline{\Gamma}^{F'}_{B'\mu} = \frac{1}{2} \sigma^{AF''}_\nu \left(\Gamma^\nu_{\mu\rho} \sigma^\rho_{AB'} + \partial_\mu \sigma^\nu_{AB'} \right). \tag{5}$$

5.5. PROBLEMS

5.6 Write the ordinary tensor affine connections (see Chapter 6) in terms of the spinor connections and the matrices σ^α and their derivatives.

Solution: From the vanishing of the covariant derivatives of the matrices $\sigma_\alpha^{AB'}$ we obtain

$$\partial_\mu \sigma_\alpha^{AB'} - \Gamma^\beta_{\alpha\mu} \sigma_\beta^{AB'} + \Gamma^A_{C\mu} \sigma_\alpha^{CB'} + \overline{\Gamma}^{B'}_{C'\mu} \sigma_\alpha^{AC'} = 0. \qquad (1)$$

Multiplying this equation by $\sigma^\rho_{AB'}$ and summing up over the indices AB' then yields

$$4\Gamma^\rho_{\alpha\mu} = \sigma^\rho_{AB'} \partial_\mu \sigma_\alpha^{AB'} + \Gamma^A_{C\mu} \sigma_\alpha^{CB'} \sigma^\rho_{AB'} + \overline{\Gamma}^{A'}_{C'\mu} \overline{\sigma}_\alpha^{C'B} \overline{\sigma}^\rho_{A'B}. \qquad (2)$$

In Eq. (2) use has been made of the fact that the matrices σ^α are Hermitian. The first term on the right-hand side of the above equation is real since

$$\overline{\sigma^\rho_{AB'} \partial_\mu \sigma_\alpha^{AB'}} = \overline{\sigma}^\rho_{A'B} \partial_\mu \overline{\sigma}_\alpha^{A'B} = \sigma^\rho_{BA'} \partial_\mu \sigma_\alpha^{BA'} = \sigma^\rho_{AB'} \partial_\mu \sigma_\alpha^{AB'}. \qquad (3)$$

Hence we finally obtain

$$4\Gamma^\rho_{\alpha\mu} = \frac{1}{2} \sigma^\rho_{AB'} \partial_\mu \sigma_\alpha^{AB'} + \Gamma^A_{C\mu} \sigma_\alpha^{CB'} \sigma^\rho_{AB'} + \text{complex conjugate}. \qquad (4)$$

5.7 Find the relationship between the spinor affine conections $\Gamma^B_{A\mu}$ and the spin affine connections Γ^A_{BC}.

Solution: The solution is left for the reader.

5.8 Show that if $f_{\alpha\beta}$ is the electromagnetic field tensor and $^*f_{\alpha\beta}$ is its dual, then the spinor equivalent to the dual is given by Eq. (5.86).

Solution: The spinor equivalent to the dual $^*f_{\alpha\beta}$ is given by

$$^*f_{AB'CD'} = \sigma^\alpha_{AB'} \sigma^\beta_{CD'} {}^*f_{\alpha\beta}$$

$$= \frac{1}{2} \sigma^\alpha_{AB'} \sigma^\beta_{CD'} \sqrt{-g} \epsilon_{\alpha\beta\gamma\delta} g^{\gamma\mu} g^{\delta\nu} f_{\mu\nu} = \frac{1}{2} \sigma^\alpha_{AB'} \sigma^\beta_{CD'} \epsilon^{\mu\nu}_{\alpha\beta} f_{\mu\nu}. \qquad (1)$$

Changing now the spacetime tensorial indices in Eq. (1) by the spinorial indices, we get

$$^*f_{AB'CD'} = \frac{1}{2} \epsilon^{KL'MN'}_{AB'CD'} f_{KL'MN'}. \qquad (2)$$

Using now Eqs. (5) and (6) of Problem 5.3 in the above formula, we obtain

$$^*f_{AB'CD'} = \frac{i}{2} (f_{AD'CB'} - f_{CB'AD'}) = i f_{AD'CB'}. \qquad (3)$$

Hence we obtain, using (5.83),

$$^*f_{AB'CD'} = i\left(\epsilon_{AC}\overline{\phi}_{D'B'} + \phi_{AC}\epsilon_{D'B'}\right)$$

or

$$^*f_{AB'CD'} = i\left(\epsilon_{AC}\overline{\phi}_{B'D'} - \phi_{AC}\epsilon_{B'D'}\right), \quad (4)$$

where use has been made of the fact that ϕ_{AB} is symmetric and $\epsilon_{B'D'}$ is skew-symmetric.

5.9 Find the spinor equivalent to the tensor

$$f^+_{\mu\nu} = f_{\mu\nu} + i\,{}^*f_{\mu\nu}. \quad (1)$$

Solution: Using Eqs. (5.83) and (5.85) we obtain

$$f^+_{AB'CD'} = 2\phi_{AC}\epsilon_{B'D'}. \quad (2)$$

5.10 Find the expression of the spinor equivalent to the energy-momentum tensor of the electromagnetic field.

Solution: The energy-momentum tensor of the electromagnetic field is given by Eq. (62) of Appendix A, Chapter 6:

$$T_{\mu\nu} = \frac{1}{4\pi}\left(\frac{1}{4}g_{\mu\nu}f_{\alpha\beta}f^{\alpha\beta} - f_{\mu\alpha}f_\nu{}^\alpha\right). \quad (1)$$

Using spinor notation, the above expression may subsequently be written in the form

$$T_{AB'CD'} = \frac{1}{4\pi}\left(\frac{1}{4}\epsilon_{AC}\epsilon_{B'D'}f_{EF'GH'}f^{EF'GH'} - f_{AB'EF'}f_{CD'}{}^{EF'}\right). \quad (2)$$

The two quadratic terms in f in Eq. (2) may be calculated. We then obtain

$$f_{EF'GH'}f^{EF'GH'} = 2\left(\phi_{EG}\phi^{EG} + \overline{\phi}_{F'H'}\overline{\phi}^{F'H'}\right), \quad (3)$$

$$f_{AB'EF'}f_{CD'}{}^{EF'} = -2\phi_{AC}\overline{\phi}_{B'D'} + \epsilon_{AC}\overline{\phi}_{B'F'}\overline{\phi}_{D'}{}^{F'} + \phi_{AE}\phi_C{}^E\epsilon_{B'D'}. \quad (4)$$

Denoting now the spinor equivalent to the energy-momentum tensor $T_{AB'CD'}$ by T_{mn}, where

$$T_{mn} = T_{A+C,B'+D'}, \quad (5)$$

5.5. PROBLEMS

with $m, n = 0, 1, 2$, and using the notation given by Eq. (5.84) for the electromagnetic field spinor ϕ_{AB} we then obtain from Eq. (2) the following simple formula:

$$T_{mn} = \frac{1}{2\pi} \phi_m \bar{\phi}_n. \tag{6}$$

Here $m, n = 0, 1, 2$.

5.11 Show that the energy-momentum tensor for the neutrino field is given by

$$T_{\mu\nu} = i(\sigma_\mu^{AX'} \eta_A \nabla_\nu \bar{\eta}_{X'} + \sigma_\nu^{AX'} \eta_A \nabla_\mu \bar{\eta}_{X'}$$
$$- \sigma_\mu^{AX'} \bar{\eta}_{X'} \nabla_\nu \eta_A - \sigma_\nu^{AX'} \bar{\eta}_{X'} \nabla_\mu \eta_A). \tag{1}$$

Solution: The solution is left for the reader.

5.12 The "standard" form of the Dirac equation is given by

$$\gamma^\mu (\nabla_\mu - ieA_\mu) \psi = m\psi, \tag{1}$$

in units in which $\hbar = c = 1$, where ψ is a four-component spinor, γ^μ are four 4×4 matrices, and m and e are the mass and charge of the particle. Show that ψ can be expressed in terms of the two 2-component spinors α^A and β^A, as

$$\psi = \begin{pmatrix} \alpha_0 \\ \alpha_1 \\ \bar{\beta}^{\dot{0}} \\ \bar{\beta}^{\dot{1}} \end{pmatrix}. \tag{2}$$

Finally, show that Eq. (1) yields the two equations (5.96). Find the matrices γ^μ and write them in terms of the three Pauli matrices and the 2×2 unit matrix.

Solution: The solution is left for the reader.

5.13 "Ghost neutrinos" in general relativity are solutions of the Einstein-Dirac equations with a neutrino current but with zero energy-momentum. Solve the Einstein field equations for a static, plane-symmetric spacetime generated by neutrinos. [See T.M. Davis and J.R. Ray, *Phys. Rev. D* **9**, 331-333 (1974); also C. Collinson and P. Morris, *Nuovo Cimento* **16B**, 273 (1973).]

Solution: The Einstein-Dirac equations are given by

$$G_{\mu\nu} = R_{\mu\nu} - \frac{1}{2} g_{\mu\nu} R = \frac{8\pi G}{c^4} T_{\mu\nu}, \tag{1}$$

$$T_{\mu\nu} = -\frac{\hbar c}{4}\left(\psi^\dagger \gamma_\mu \psi_{;\nu} - \psi^\dagger_{;\nu}\gamma_\mu\psi + \psi^\dagger\gamma_\nu\psi_{;\mu} - \psi^\dagger_{;\mu}\gamma_\nu\psi\right), \qquad (2)$$

where ψ satisfies the zero-mass Dirac equation

$$\gamma^\alpha \psi_{;\alpha} = 0, \qquad (3)$$

and the semicolon denotes covariant differentiation. Since for zero-mass Dirac particles the trace of the energy-momentum tensor vanishes, the scalar curvature R also vanishes, and the Einstein equations reduce to

$$R_{\mu\nu} = \frac{8\pi G}{c^4} T_{\mu\nu}. \qquad (4)$$

Spacetimes with plane symmetry along the x axis can be given by

$$ds^2 = e^{2u}\left(dt^2 - dx^2\right) - e^{2v}\left(dy^2 - dz^2\right), \qquad (5)$$

where u and v are functions of x and t only, and the speed of light c is taken as unity. For the static case u, v, ψ, etc. depend on only x.

The exact solution to the Einstein-Dirac equations is given by

$$u = \ln(kx+1)^{-1/4}, \qquad v = \ln(kx+1)^{1/2}, \qquad (6)$$

and

$$\psi_a = (kx+1)^{-3/8}\psi_{a0}, \qquad (7)$$

where

$$\psi_{a0} = a\begin{pmatrix} 1 \\ \pm 1 \\ i \\ \pm i \end{pmatrix}, \qquad (8)$$

and k and a are arbitrary constants. The neutrino current density,

$$J^\mu = i\psi^\dagger \gamma^\mu \psi, \qquad (9)$$

is consequently given by

$$J^\mu = 4\,|\,a\,|^2\,(kx+1)^{-3/4}\,(\delta_0^\mu \pm \delta_1^\mu). \qquad (10)$$

Equations (6) and (7) represent an exact solution to the Einstein-Dirac equations for a static, plane-symmetric spacetime. Its most interesting property is that the neutrino energy-momentum tensor vanishes, whereas the neutrino field and current density do not vanish.

5.14 Show that the solution to the Einstein-Dirac equations given in Problem 5.13 can be generalized to represent "non-ghost" neutrinos,

$$u = -\frac{1}{4}\ln(ax+b) + cx + d, \tag{1}$$

$$v = \frac{1}{2}\ln(ax+b), \tag{2}$$

$$\psi = e^{-(v+u/2)}\left(\cos\omega x + i\gamma^1\gamma^0 \sin\omega x\right)e^{-i\omega t}\begin{pmatrix} s \\ \pm s \\ q \\ \pm q \end{pmatrix}e^{i\phi}, \tag{3}$$

where a, b are constants, and s, q and ϕ are arbitrary real numbers. All components of $T_{\mu\nu}$ are zero except T_{00} and T_{11}, which are equal.

Solution: The solution is left for the reader. [See K.R. Rechenick and J.M. Cohen, *Phys. Rev.* D **19**, 1635-1640 (1979).]

5.15 Generalize the solutions given in Problems **5.13** and **5.14** to the case in which an electromagnetic field is present. Show that in the presence of an electromagnetic field even the time-dependent Dirac field has ghost solutions, but the solutions become ghost-free in the presence of charged matter. Show also that the time-independent Dirac field has a ghost-free solution if the neutrinos are considered to possess some mass.

Solution: The solution is left for the reader. [See K.D. Krori, T. Chaudhury and R. Bhattacharjee, *Phys. Rev.* D **25**, 1492-1498 (1982).]

5.6 References for Further Reading

W.L. Bade and J. Jehle, An introduction to spinors, *Revs. Mod. Phys.* **25**, 714-728 (1953). (Section 5.2)

J.D. Bjorken and D. Drell, *Relativistic Quantum Mechanics* (McGraw-Hill Book Co., New York, 1964). (Section 5.4)

M. Carmeli, *Group Theory and General Relativity* (McGraw-Hill, New York, 1977).

M. Carmeli, *Classical Fields: General Relativity and Gauge Theory* (John Wiley, 1982).

M. Carmeli, E. Leibowitz and N. Nissani, *Gravitation: SL(2,C) Gauge Theory and Conservation Laws* (World Scientific, Singapore, 1990). (Sections

5.2-5.4)

E.M. Corson, *Introduction to Tensors, Spinors and Relativistic Wave Equations* (Hafner Publishing Co., New York, 1953). (Section 5.4)

P.A.M. Dirac, *Proc. Roy. Soc. A* **117**, 610 (1928). (Sections 5.2-5.5)

M. Fierz, Uber die Relativistische Theorie kraftefreier Teilchen mit beliebigen Spin, *Helv. Phys. Acta* **12**, 3 (1938). (Sections 5.2-5.5)

L. Infeld and B.L. van der Waerden, The wave equation of the electron in the general relativity theory, *Sb. preuss. Akad. Wiss., Phys.-mat.* Kl. **380** (1933). (Section 5.2)

J.D. Jackson, *Classical Electrodynamics* (John Wiley, New York, 1975). (Section 5.1)

T. Kahan (Editor), *Theory of Groups in Classical and Quantum Theory*, Vol. I (Oliver and Boyd, London, 1965). (Sections 5.2-5.5)

W. Pauli, Zur Quantenmechanik des magnetischen Electrons, *Z. Physik* **43**, 601 (1927). (Sections 5.2-5.5)

R. Penrose, A spinor approach to general relativity, *Ann. Phys. (N.Y.)* **10**, 171-201 (1960). (Sections 5.2-5.4)

O. Veblen and J. von Neumann, *Geometry of Complex Dynamics* (The Institute for Advanced Study, Princeton, New Jersey, 1958). (Section 5.2)

B.L. van der Waerden, *Die gruppentheoretische Methode in der Quantenmechanik* (Springer, Berlin, 1932). (Sections 5.2-5.5)

Chapter 6

The Gravitational Field Spinors

In this chapter the gravitational field spinors are presented. These include the curvature spinor and the spinors equivalent to the Riemann, Weyl, Ricci and Einstein tensors. A decomposition of the Riemann spinor into its irreducible components is also given. We start the chapter with a brief review of the essentials of general relativity theory.

6.1 Elements of General Relativity

In this section a brief review of general relativity theory is given. Only the essentials of the theory that is needed for the chapter are given. We begin the discussion with a brief review of Riemannian geometry, followed by a description of the physical foundations of general relativity. These are the principles of equivalence and of general covariance. The gravitational field equations are then derived in a tensorial form. The Schwarzschild solution of Einstein's field equations is derived. Experimental verification of general relativity is subsequently given. The section is then concluded with a review of the problem of motion in the gravitational field.

6.1.1 Riemannian Geometry

Transformation of Coordinates

Any four independent variables x^μ, where Greek letters take the values 0, 1, 2, 3, may be considered as the coordinates of a four-dimensional space V_4. Each set of values of x^μ defines a point of V_4. Let there be another set of coordinates x'^μ related to the first set x^ν by

$$x'^\mu = f^\mu(x^\nu), \tag{6.1}$$

where f^μ are four independent real functions of x^ν. A necessary and sufficient condition that f^μ be independent is that their Jacobian,

$$\left| \frac{\partial f^\mu}{\partial x^\nu} \right| = \begin{vmatrix} \frac{\partial f^0}{\partial x^0} & \cdots & \frac{\partial f^3}{\partial x^0} \\ \vdots & & \\ \frac{\partial f^0}{\partial x^3} & \cdots & \frac{\partial f^3}{\partial x^3} \end{vmatrix}, \tag{6.2}$$

does not vanish identically. Equation (6.1) defines a *transformation of coordinates* in the space V_4. When the Jacobian is different from zero, one can also write x^μ in terms of x'^ν as

$$x^\mu = g^\mu(x'^\nu). \tag{6.3}$$

A *direction* at a point P in the space V_4 is determined by the differential dx^μ. The same direction is determined in another set of coordinates x'^μ by the differential dx'^μ. The two differentials are related, using Eq. (6.1), by

$$dx'^\mu = \frac{\partial x'^\mu}{\partial x^\nu} dx^\nu = \frac{\partial f^\mu}{\partial x^\nu} dx^\nu. \tag{6.4}$$

Here the Einstein summation convention is used, according to which repeated Greek indices are summed over the values 0, 1, 2, 3.

Contravariant Vectors

Let two sets of functions V^μ and V'^μ be related by

$$V'^\mu = \frac{\partial x'^\mu}{\partial x^\nu} V^\nu, \tag{6.5}$$

similar to the way the differentials dx'^μ and dx^μ are related. V^μ and V'^μ are then called the *components* of a *contravariant vector* in the coordinate

6.1. ELEMENTS OF GENERAL RELATIVITY

systems x^μ and x'^μ, respectively. Hence any four functions of the x's in one coordinate system can be taken as the components of a contravariant vector whose components in any other coordinate system are given by Eq. (6.5).

A contravariant vector determines a direction at each point of the space V_4. Let V^μ be the components of a contravariant vector and let dx^μ be a displacement in the direction of V^μ. Then $dx^0/V^0 = \cdots = dx^3/V^3$. This set of equations admits three independent $f^k(x^\mu) = c^k$, where $k = 0, 1, 2$, and the c's are arbitrary constants and the matrix $\partial f^k/\partial c^\mu$ is of rank three. The functions f^k are solutions of the partial differential equation $V^\nu \partial f^k/\partial x^\nu = 0$. Hence using the transformation laws (6.1) and (6.3) one obtains $V'^k = 0$ for $k = 0, 1, 2$, and $V'^3 \neq 0$. Hence a system of coordinates can be chosen in terms of which all components but one of a given contravariant vector are equal to zero.

Invariants. Covariant Vectors

Two functions $f(x)$ and $f'(x')$ define an *invariant* if they are reducible to each other by a coordinate transformation.

Let f be a function of the coordinates. Then

$$\frac{\partial f}{\partial x'^\mu} = \frac{\partial f}{\partial x^\nu} \frac{\partial x^\nu}{\partial x'^\mu}. \tag{6.6}$$

Two sets of functions V_μ and V'_μ are called the components of a *covariant vector* in the systems x and x', respectively, if they are related by the transformation law of the form (6.6),

$$V'_\mu = \frac{\partial x^\nu}{\partial x'^\mu} V_\nu. \tag{6.7}$$

For example, if f is a scalar function, then $\partial f/\partial x^\mu$ is a covariant vector. It is called the *gradient* of f. The product $V^\mu W_\mu$ is an invariant if V is a contravariant vector and W is a covariant vector. Conversely, if the quantity $V^\mu W_\mu$ is an invariant and either V^μ or W_μ are arbitrary vectors, then the other set is a vector.

Tensors

Tensors of any order are defined by generalizing Eqs. (6.5) and (6.7). Thus the equation

$$T'^{\mu_1\cdots\mu_m}_{\nu_1\cdots\nu_n} = \frac{\partial x'^{\mu_1}}{\partial x^{\rho_1}} \cdots \frac{\partial x'^{\mu_m}}{\partial x^{\rho_m}} \frac{\partial x^{\sigma_1}}{\partial x'^{\nu_1}} \cdots \frac{\partial x^{\sigma_n}}{\partial x'^{\nu_n}} T^{\rho_1\cdots\rho_m}_{\sigma_1\cdots\sigma_n} \tag{6.8}$$

defines a mixed tensor of order $m + n$, contravariant of the mth order and covariant of the nth order. If the Kronecker delta function is taken as the components of a mixed tensor of the second order in one set of coordinates, for example, then it defines the components of a tensor in any set of coordinates. An invariant is a tensor of zero order and a vector is a tensor of order one.

When the relative position of two indices, either contravariant or covariant, is immaterial, the tensor is called *symmetric* with respect to these indices. When the relative position of two indices of a tensor is interchanged and the tensor obtained differs only in sign from the original one, the tensor is called *skew-symmetric* with respect to these indices. The process by means of which from a mixed tensor of order r one obtains a tensor of order $r - 2$ is called *contraction*.

Metric Tensor

Let $g_{\mu\nu}$ be the components of the *metric tensor*, i.e., a symmetric covariant tensor, which is a function of coordinates, and let $g = \det g_{\mu\nu}$. The quantity $g^{\mu\nu}$, denoting the cofactor of $g_{\mu\nu}$ divided by g, is a symmetric contravariant tensor and satisfies

$$g^{\mu\rho} g_{\nu\rho} = \delta^{\mu}_{\nu}. \tag{6.9}$$

The element of length is defined by means of a quadratic differential form $ds^2 = g_{\mu\nu} dx^{\mu} dx^{\nu}$. By means of the tensors $g_{\mu\nu}$ and $g^{\mu\nu}$ one can lower or raise tensor indices:

$$T^{\mu}{}_{\nu\rho} = g^{\mu\sigma} T_{\sigma\nu\rho}, \tag{6.10a}$$

$$T_{\alpha}{}^{\beta\gamma} = g_{\alpha\rho} T^{\rho\beta\gamma}. \tag{6.10b}$$

Certain other quantities transform according to the law

$$T'^{\mu\cdots}_{\alpha\cdots} = J^N \frac{\partial x'^{\mu}}{\partial x^{\rho}} \frac{\partial x^{\beta}}{\partial x'^{\alpha}} \cdots T^{\rho\cdots}_{\beta\cdots}. \tag{6.11}$$

Here J is the Jacobian determinant $|\partial x^{\alpha}/\partial x'^{\beta}|$. The superscript N is the power to which J is raised. $T^{\mu\cdots}_{\nu\cdots}$ is called a *tensor density* of weight N. For example, if g' denotes $\det g'_{\mu\nu}$ then $g' = J^2 g$, where $g = \det g_{\mu\nu}$. Hence one has for the four-dimensional elements in two coordinate systems the equality:

$$\sqrt{-g}\, d^4 x = \sqrt{-g'}\, d^4 x'. \tag{6.12}$$

6.1. ELEMENTS OF GENERAL RELATIVITY

Christoffel Symbols

From the two tensors $g_{\mu\nu}$ and $g^{\mu\nu}$ one can define the two functions

$$\Gamma_{\alpha\rho\sigma} = \frac{1}{2}\left(\frac{\partial g_{\rho\alpha}}{\partial x^\sigma} + \frac{\partial g_{\sigma\alpha}}{\partial x^\rho} - \frac{\partial g_{\rho\sigma}}{\partial x^\alpha}\right), \qquad (6.13)$$

$$\Gamma^\mu_{\rho\sigma} = g^{\mu\alpha}\Gamma_{\alpha\rho\sigma}. \qquad (6.14)$$

They are symmetric in ρ and σ, and are called the *Christoffel symbols* of the *first* and *second* kind, respectively.

Both kinds of Christoffel symbols are not components of tensors. By starting with the differential transformation law for $g_{\mu\nu}$ it is not too difficult to show that $\Gamma_{\alpha\rho\sigma}$ transforms according to the following relation (see Problem 6.10):

$$\Gamma'_{\nu\mu\alpha} = \frac{\partial x^\beta}{\partial x'^\mu}\frac{\partial x^\gamma}{\partial x'^\alpha}\frac{\partial x^\delta}{\partial x'^\nu}\Gamma_{\delta\beta\gamma} + g_{\beta\gamma}\frac{\partial x^\beta}{\partial x'^\nu}\frac{\partial^2 x^\gamma}{\partial x'^\mu \partial x'^\alpha}. \qquad (6.15)$$

Making use of the transformation law for $g^{\alpha\beta}$ then leads to the transformation law of $\Gamma^\delta_{\beta\nu}$ as

$$\Gamma'^\delta_{\beta\nu} = \frac{\partial x'^\delta}{\partial x^\alpha}\frac{\partial x^\mu}{\partial x'^\beta}\frac{\partial x^\sigma}{\partial x'^\nu}\Gamma^\alpha_{\mu\sigma} + \frac{\partial x'^\delta}{\partial x^\sigma}\frac{\partial^2 x^\sigma}{\partial x'^\beta x'^\nu}. \qquad (6.16)$$

From Eq. (6.13) we obtain

$$\Gamma^\mu_{\alpha\mu} = \frac{1}{2}g^{\mu\nu}\frac{\partial g_{\mu\nu}}{\partial x^\alpha}. \qquad (6.17)$$

This equation can be rewritten in terms of the determinant g of $g_{\mu\nu}$. The rule for expansion of a determinant leads to the formula

$$\frac{\partial g}{\partial g_{\mu\nu}} = \Delta^{\mu\nu}, \qquad (6.18)$$

where $\Delta^{\mu\nu}$ is the cofactor of the element $g_{\mu\nu}$. From the law for obtaining the inverse of a determinant, and from the definition of $g^{\mu\nu}$, Eq. (6.18) may be written as

$$\frac{\partial g}{\partial g_{\mu\nu}} = gg^{\mu\nu}, \qquad (6.19)$$

and consequently

$$dg = gg^{\mu\nu}dg_{\mu\nu} = -gg_{\mu\nu}dg^{\mu\nu}. \qquad (6.20)$$

Hence we have
$$\partial_\alpha g = g g^{\mu\nu} \partial_\alpha g_{\mu\nu} = -g g_{\mu\nu} \partial_\alpha g^{\mu\nu}. \tag{6.21}$$

The use of Eq. (6.21) enables us to write Eq. (6.17) in the form
$$\Gamma^\mu_{\alpha\mu} = \frac{1}{2g}\frac{\partial g}{\partial x^\alpha} = \frac{1}{\sqrt{-g}}\frac{\partial \sqrt{-g}}{\partial x^\alpha}. \tag{6.22}$$

Covariant Differentiation

We have seen that the derivatives of an invariant are the components of a covariant vector. This is the only case for a general system of coordinates in which the derivative of a tensor is a tensor. However, there are expressions involving first derivatives which are components of a tensor. To see this we proceed as follows.

Let V^μ and V'^ν be a contravariant vector in two coordinate systems x and x'. Then
$$V^\mu = V'^\nu \frac{\partial x^\mu}{\partial x'^\nu}. \tag{6.23}$$

Differentiating this equation with respect to x^α and using Eq. (6.16) gives (see Problem 6.11):
$$\frac{\partial V^\mu}{\partial x^\alpha} = \left(\frac{\partial V'^\rho}{\partial x'^\nu} + V'^\sigma \Gamma'^\rho_{\sigma\nu}\right) \frac{\partial x'^\nu}{\partial x^\alpha} \frac{\partial x^\mu}{\partial x'^\rho} - V^\rho \Gamma^\mu_{\rho\alpha}. \tag{6.24}$$

Hence if we define a *covariant derivative* of V^μ by
$$\nabla_\alpha V^\mu = \partial_\alpha V^\mu + \Gamma^\mu_{\rho\alpha} V^\rho, \tag{6.25}$$

Eq. (6.24) can be written as
$$\nabla_\alpha V^\mu = \nabla_\nu V'^\rho \frac{\partial x'^\nu}{\partial x^\alpha} \frac{\partial x^\mu}{\partial x'^\rho}. \tag{6.26}$$

Therefore $\nabla_\alpha V^\mu$ is a mixed tensor of second order.

In the same way one shows that the covariant derivative of a covariant vector V_μ is given by:
$$\nabla_\alpha V_\mu = \partial_\alpha V_\mu - \Gamma^\rho_{\mu\alpha} V_\rho, \tag{6.27}$$

From the above equation one has for the *curl* of a vector V_μ:
$$\nabla_\beta V_\alpha - \nabla_\alpha V_\beta = \partial_\beta V_\alpha - \partial_\alpha V_\beta. \tag{6.28}$$

6.1. ELEMENTS OF GENERAL RELATIVITY

Hence a necessary and sufficient condition that the first covariant derivative of a covariant vector be symmetric is that the vector be a gradient.

It is easily seen, using the law of covariant differentiation of tensors (see Problem 6.12), that

$$\nabla_\rho g^{\mu\nu} = 0, \tag{6.29a}$$

$$\nabla_\rho g_{\mu\nu} = 0, \tag{6.29b}$$

$$\nabla_\rho \delta^\mu_\nu = 0. \tag{6.29c}$$

Other properties of covariant differentiation can be established (see Problem 6.13).

Riemann and Ricci Tensors

If we differentiate covariantly the tensor $\nabla_\alpha V_\mu$, given by Eq. (6.27), we obtain

$$(\nabla_\gamma \nabla_\beta - \nabla_\beta \nabla_\gamma) V_\alpha = R^\delta{}_{\alpha\beta\gamma} V_\delta, \tag{6.30}$$

where $R^\delta{}_{\alpha\beta\gamma}$ is called the *Riemann tensor* and is given by

$$R^\delta{}_{\alpha\beta\gamma} = \partial_\beta \Gamma^\delta_{\alpha\gamma} - \partial_\gamma \Gamma^\delta_{\alpha\beta} + \Gamma^\mu_{\alpha\gamma} \Gamma^\delta_{\mu\beta} - \Gamma^\mu_{\alpha\beta} \Gamma^\delta_{\mu\gamma}. \tag{6.31}$$

A generalization of Eq. (6.30) to an arbitrary tensor can be made (see Problem 6.14).

One can show that in order that there can exist a coordinate system in which the first covariant derivatives reduce to ordinary ones at every point in space, it is necessary and sufficient that the Riemann tensor be zero and that the coordinates be those in which the metric is constant.

One notices that the Riemann tensor satisfies

$$R_{\alpha\beta\gamma\delta} = -R_{\beta\alpha\gamma\delta} = -R_{\alpha\beta\delta\gamma} = R_{\gamma\delta\alpha\beta}, \tag{6.32a}$$

$$R_{\alpha\beta\gamma\delta} + R_{\alpha\gamma\delta\beta} + R_{\alpha\delta\beta\gamma} = 0. \tag{6.32b}$$

Moreover, counting the number of components, one finds that in a four-dimensional space the Riemann tensor has 20 components.

From the Riemann tensor one can define the *Ricci tensor* and the *Ricci scalar* by

$$R_{\mu\nu} = R^\alpha{}_{\mu\alpha\nu} = \frac{1}{\sqrt{-g}} \left(\sqrt{-g}\,\Gamma^\alpha_{\mu\nu}\right)_{,\alpha} - \left(\ln \sqrt{-g}\right)_{,\mu\nu} - \Gamma^\alpha_{\mu\beta} \Gamma^\beta_{\nu\alpha}, \tag{6.33}$$

$$R = R^\mu{}_\mu, \tag{6.34}$$

respectively. Here a comma denotes partial differentiation, $f_{,\alpha} = \partial_\alpha f$. The *Einstein tensor* is then defined by

$$G_{\mu\nu} = R_{\mu\nu} - \frac{1}{2} g_{\mu\nu} R. \tag{6.35}$$

The last important tensor constructed from the Riemann tensor is the *Weyl conformal tensor* (see Section 6.1.8):

$$C_{\rho\sigma\mu\nu} = R_{\rho\sigma\mu\nu} - \frac{1}{2} \left(g_{\rho\mu} R_{\nu\sigma} - g_{\rho\nu} R_{\mu\sigma} - g_{\sigma\mu} R_{\nu\rho} + g_{\sigma\nu} R_{\mu\rho} \right)$$

$$- \frac{1}{6} \left(g_{\rho\nu} g_{\mu\sigma} - g_{\rho\mu} g_{\nu\sigma} \right) R. \tag{6.36}$$

It has the special property that

$$C^\rho{}_{\mu\rho\nu} = 0. \tag{6.37}$$

Furthermore, if the Weyl tensor vanishes everywhere, then the metric is said to be *conformally flat*. (Two spaces V and $\tilde{V}$ are called conformal spaces if their metric tensors $g_{\mu\nu}$ and $\tilde{g}_{\mu\nu}$ are related by $\tilde{g}_{\mu\nu} = e^\beta g_{\mu\nu}$, where β is a function of the coordinates.) That is, there exists a mapping such that $g_{\mu\nu}$ can be diagonalized, with $\pm\beta(x)$ appearing in the diagonal positions, and where $\beta(x)$ is some function. This follows from the fact that the Weyl tensor can be expressed entirely in terms of the density $\tilde{g}_{\mu\nu} = (-g)^{-1/4} g_{\mu\nu}$ and its inverse, and is equal to the Riemann tensor formed by replacing $g_{\mu\nu}$ by $\tilde{g}_{\mu\nu}$, $R_{\alpha\beta\gamma\delta}(\tilde{g}_{\mu\nu}) = C_{\alpha\beta\gamma\delta}(g_{\mu\nu})$.

Consequently, the vanishing of the Weyl tensor implies the vanishing of $R_{\alpha\beta\gamma\delta}(\tilde{g}_{\mu\nu})$, which in turn implies that there exists a mapping such that $\tilde{g}_{\mu\nu}$ is everywhere diagonal, with ± 1 appearing along the diagonal. Only g is arbitrary and $\pm(-g)^{1/4}$ appears along the diagonal of $g_{\mu\nu}$.

Geodesics

The differential equations of the curves of extremal length are called *geodesic equations*. To find their equations we seek the relations which must be satisfied to give a stationary value to the integral $\int ds$. Hence we have to find the solution of the variational problem

$$\delta \int L \, ds = 0, \tag{6.38}$$

6.1. ELEMENTS OF GENERAL RELATIVITY

where the Lagrangian L is given by

$$L = \left(g_{\mu\nu} \frac{dx^\mu}{ds} \frac{dx^\nu}{ds} \right)^{1/2}. \tag{6.39}$$

Accordingly we have

$$\delta \int L\, ds = \int \left[\frac{\partial L}{\partial x^\mu} \delta x^\mu + \frac{\partial L}{\partial (dx^\mu/ds)} \delta \left(\frac{dx^\mu}{ds} \right) \right] ds. \tag{6.40}$$

The second term of the integrand may be written as the two terms

$$\frac{d}{ds} \left[\frac{\partial L}{\partial (dx^\mu/ds)} \delta x^\mu \right] - \frac{d}{ds} \left[\frac{\partial L}{\partial (dx^\mu/ds)} \right] \delta x^\mu. \tag{6.41}$$

On integration, the first of these expressions contributes nothing since the variations are assumed to vanish at the end points of the curve.

As expected, the equation obtained is the usual Lagrange equation:

$$\frac{d}{ds} \frac{\partial L}{\partial (dx^\mu/ds)} - \frac{\partial L}{\partial x^\mu} = 0. \tag{6.42}$$

A simple calculation then gives, using the Lagrangian given by Eq. (6.39),

$$\frac{d^2 x^\mu}{ds^2} + \Gamma^\mu_{\alpha\beta} \frac{dx^\alpha}{ds} \frac{dx^\beta}{ds} = 0. \tag{6.43}$$

Bianchi Identities

A study of Eq. (6.16) shows that it is always possible to choose a coordinate system in which all the Christoffel symbols vanish at a point. For, suppose the Christoffel symbols do not vanish at a point A. One can then carry out the coordinate transformation

$$x'^\alpha = x^\alpha - x^\alpha{}_A + \frac{1}{2} \Gamma^\alpha_{\beta\gamma}(A) \left(x^\beta - x^\beta{}_A \right) \left(x^\gamma - x^\gamma{}_A \right), \tag{6.44}$$

where the subscript A indicates to the value at the point A. By Eq. (6.16) one finds that the Christoffel symbols in the new coordinate system vanish at the point A.

A coordinate system for which the Christoffel symbols vanish at a point is called *geodesic*. (It is also possible to transform away the Christoffel symbols along a given curve.)

If we choose a geodesic coordinate system at a point A, then at A one has

$$\nabla_\nu R^\mu{}_{\delta\beta\gamma} = \partial_\beta \partial_\nu \Gamma^\mu_{\delta\gamma} - \partial_\gamma \partial_\nu \Gamma^\mu_{\delta\beta}. \tag{6.45}$$

Consequently, at the point A one has:

$$\nabla_\nu R^\mu{}_{\delta\beta\gamma} + \nabla_\gamma R^\mu{}_{\delta\nu\beta} + \nabla_\beta R^\mu{}_{\delta\gamma\nu} = 0. \tag{6.46}$$

Since the terms of this equation are components of a tensor, this equation holds for any coordinate system and at each point. Hence Eq. (6.46) is an identity throughout the space. It is known as the *Bianchi identities*.

Multiplication of Eq. (6.46) by $g^{\delta\beta} \delta^\gamma_\mu$ gives

$$g^{\delta\beta} \left(\nabla_\nu R^\gamma{}_{\delta\beta\gamma} + \nabla_\gamma R^\gamma{}_{\delta\nu\beta} + \nabla_\beta R^\gamma{}_{\delta\gamma\nu} \right) = 0. \tag{6.47}$$

Using the symmetry properties of the Riemann tensor, the last equation becomes:

$$\nabla_\nu \left(R_\gamma{}^\nu - \frac{1}{2} \delta^\nu_\gamma R \right) = \nabla_\nu G_\gamma{}^\nu = 0. \tag{6.48}$$

Equation (6.48) is called the *contracted Bianchi identity*.

After having developed the mathematical tools to describe general relativity theory, we now turn to the physical foundations of the theory.

6.1.2 Principle of Equivalence

Null Experiments. Eötvös Experiment

One of the most interesting *null experiments* in physics is due to Eötvös, first performed in 1890 and recently repeated by Dicke. The experiment showed, in great precision, that all bodies fall with the same acceleration. The roots of the experiment go back to Newton and Galileo, who demonstrated experimentally that the gravitational acceleration of a body is independent of its composition.

The importance of the Eötvös experiment is in the fact that the null result of the experiment is a *necessary* condition for the theory of general relativity to be valid.

Eötvös employed a static torsion balance, balancing a component of the Earth's gravitational pull on the weight against the centrifugal force field of the Earth acting on the weight. He employed a horizontal torsion beam, 40 cm long, suspended by a fine wire. From the ends of the torsion beam were suspended two masses of different compositions, one lower than the

6.1. ELEMENTS OF GENERAL RELATIVITY

other. A lack of exact proportionality between the inertial and gravitational masses of the two bodies would then lead to a torque tending to rotate the balance. There appears to be no need for the one mass to be suspended lower than the other.

The experement of Eötvös showed, with an accuracy of a few parts in 10^9, that inertial and gravitational masses are equal.

In the experiment performed by Dicke, the gravitational acceleration toward the Sun of small gold and aluminium weights were compared and found to be equal with an accuracy of about one part in 10^{11}. Hence the necessary condition to be satisfied for the validity of general relativity theory seems to be rather satisfactory met.

The question therefore arises as to what extent is this experiment also a sufficient condition to be satisfied in order that general relativity theory be valid.

It has been emphasized by Dicke that gold and aluminium differ from each other rather greatly in several important aspects. First, the neutron to proton ratio is quite different in the two elements, varying from 1.08 in aluminium to 1.50 in gold. Second, the electrons in aluminium move with nonrelativistic velocities, but in gold the k-shell electrons have a 15 per cent increase in their masses as a result of their relativistic velocities. Third, the electromagnetic negative contribution to the binding energy of the nucleus varies as z^2 and represents $\frac{1}{2}$ per cent of the total mass of a gold atom, whereas it is negligible in aluminium. Fourth, the virtual pair field and other fields would be expected to be different in the two atoms. We thus conclude that the physical aspects of gold and aluminium differ substantially, and consequently the equality of their accelerations represents an important condition to be satisfied by any theory of gravitation.

Since the accuracy of the Eötvös experiment is great, the question arises as to whether it implies that the equivalence principle is very nearly valid. This is true in a limited sense; certain aspects of the equivalence principle are not supported in the slightest by the Eötvös experiment.

6.1.3 Principle of General Covariance

We have seen in the preceeding subsection that a gravitational field can be considered locally equivalent to an accelerated frame. This implies that the special theory of relativity (see Section 3.1) cannot be valid in an extended region where gravitational fields are present. A curved spacetime is needed and all laws of nature should be covariant under the most general coordinate transformations.

The original formulation of general relativity by Einstein was based on two principles: (1) the principle of equivalence (discussed in detail in the last subsection); and (2) the principle of general covariance.

The principle of general covariance is often stated in one of the following forms, which are not exactly equivalent:

(1) All coordinate systems are equally good for stating the laws of physics, and they should be treated on the same footing.

(2) The equations of physics should have tensorial forms.

(3) The equations of physics should have the same form in all coordinate systems.

According to the principle of general covariance, the coordinates become nothing more than a bookkeeping system to label the events. The principle is a valuable guide to deducing correct equations.

It has been pointed out that any spacetime physical law can be written in a covariant form and hence the principle of general covariance has no necessary physical consequences, and Einstein concurred with this view.

In spite of Einstein's acceptance of this objection, it appears that the principle of general covariance was introduced by Einstein as a generalization of the principle of special relativity and he often referred to it as the principle of general relativity. In fact the principle of equivalence (which necessarily leads to the introduction of a curved spacetime), plus the assumption of general covariance, is most of what is needed to generate Einstein's theory of general relativity. They lead directly to the idea that gravitation can be explained by means of Riemannian geometry. This is done in the next subsection.

6.1.4 Gravitational Field Equations

We have seen in Subsection 6.1.1 that the Riemannian geometry is characterized by a geometrical metric, i.e., a symmetric tensor $g_{\mu\nu}$ from which one can construct other quantities. Classical general relativity theory identifies this tensor as the gravitational potential. Hence in general relativity there are ten components to the gravitational potential, as compared with the single potential function in the Newtonian theory of gravitation.

Einstein's Field Equations

In trying to arrive at the desired gravitational field equations that the metric tensor has to satisfy, we are guided by the requirement that, in an appropriate limit, the theory should be reduced to the Newtonian gravitational

6.1. ELEMENTS OF GENERAL RELATIVITY

theory. In the latter theory, the gravitational potential ϕ is determined by the Poisson equation:
$$\nabla^2 \phi = 4\pi G \rho, \tag{6.49}$$
where G ($= 6.67 \times 10^{-8}$ cm^3 gm^{-1} sec^{-2}) is the Newton gravitational constant and ρ is the mass density of matter. Hence $g_{\mu\nu}$ should satisfy second order partial differential equations. The equations should then be related to the energy-momentum tensor $T_{\mu\nu}$ linearly. Such equations are
$$R_{\mu\nu} - \frac{1}{2} g_{\mu\nu} R = \kappa T_{\mu\nu}, \tag{6.50}$$
where κ is some constant to be determined. In cosmology theory, one sometimes adds an additional term, $\lambda g_{\mu\nu}$, to the left-hand side of Eq. (6.50). The constant λ is known as a cosmological constant.

But the contracted Bianchi identities, Eq. (6.48), show that the covariant divergence of the left-hand side of Eq. (6.50) vanishes. Hence
$$\nabla_\nu T_\mu{}^\nu = 0, \tag{6.51}$$
which expresses the covariant conservation of energy and momentum. The constant κ can be determined by going to the limit of weak gravitational field (see Problem 6.20). Its value is $\kappa = 8\pi G/c^4$. The constant κ is known as Einstein's gravitational constant.

Deduction of Einstein's Equations from Variational Principle

We start with the action integral
$$I = \int \sqrt{-g} \, (L_G - 2\kappa L_F) \, d^4x, \tag{6.52}$$
and demand its variation to be zero. Here L_G and L_F are the Lagrangians for the gravitational and other fields, respectively. We take $L_G = R$, where R is the Ricci scalar, $R = R_{\mu\nu} g^{\mu\nu}$.

The first part of the integral (6.52) gives
$$\delta \int \sqrt{-g} R \, d^4x = \int \sqrt{-g} g^{\mu\nu} \delta R_{\mu\nu} \, d^4x + \int R_{\mu\nu} \delta \left(\sqrt{-g} g^{\mu\nu} \right) d^4x. \tag{6.53}$$

To find $\delta R_{\mu\nu}$ we note that in a geodesic coordinate system one has
$$\delta R_{\mu\nu} = \nabla_\alpha \left(\delta \Gamma^\alpha_{\mu\nu} \right) - \nabla_\nu \left(\delta \Gamma^\alpha_{\mu\alpha} \right). \tag{6.54}$$

But the latter is a tensorial equation. Hence it is valid in all coordinate systems. Consequently, the first integral on the right-hand side of Eq. (6.53) can be written as

$$\int \sqrt{-g}\, g^{\mu\nu}\delta R_{\mu\nu}\,d^4x = \int \sqrt{-g}\,\nabla_\alpha\left(g^{\mu\nu}\delta\Gamma^\alpha_{\mu\nu} - g^{\mu\alpha}\delta\Gamma^\beta_{\mu\beta}\right)d^4x, \qquad (6.55)$$

and hence (by Problem 6.16) is equal to

$$\int \partial_\alpha\left[\sqrt{-g}\left(g^{\mu\nu}\delta\Gamma^\alpha_{\mu\nu} - g^{\mu\alpha}\delta\Gamma^\beta_{\mu\beta}\right)\right]d^4x. \qquad (6.56)$$

This integral, however, vanishes since by Gauss' theorem it is equal to a surface integral which is equal to zero in consequence of the vanishing of the variations at the boundary.

The second integral on the right-hand side of Eq. (6.53) gives, by Eq. (6.21),

$$\int R_{\mu\nu}\delta\left(\sqrt{-g}\,g^{\mu\nu}\right)d^4x = \int \sqrt{-g}\left(R_{\mu\nu} - \frac{1}{2}g_{\mu\nu}R\right)\delta g^{\mu\nu}\,d^4x. \qquad (6.57)$$

The second part of the integral (6.52) leads to (see Problem 6.21)

$$\delta\int \sqrt{-g}\,L_F\,d^4x = -\frac{1}{2}\int \sqrt{-g}\,T_{\mu\nu}\delta g^{\mu\nu}\,d^4x, \qquad (6.58)$$

where $T_{\mu\nu}$ is the energy-momentum tensor and is given by

$$T_{\mu\nu} = \frac{-2}{\sqrt{-g}}\left[\left(\frac{\partial(\sqrt{-g}\,L_F)}{\partial g^{\mu\nu}_{,\alpha}}\right)_{,\alpha} - \frac{\partial(\sqrt{-g}\,L_F)}{\partial g^{\mu\nu}}\right], \qquad (6.59)$$

and a comma denotes partial differentiation, $f_{,\alpha} = \partial_\alpha f$. Combining Eqs. (6.52), (6.57) and (6.58) then leads to the field equations (6.50):

$$R_{\mu\nu} - \frac{1}{2}g_{\mu\nu}R = \kappa T_{\mu\nu}. \qquad (6.60)$$

The Electromagnetic Energy-Momentum Tensor

The energy-momentum tensor $T_{\mu\nu}$ for the electromagnetic field is obtained from the general expression (6.59) with the field Lagrangian L_F given by the first part of the Lagrangian density (5.22) of Chapter 5, namely,

$$L_F = -\frac{1}{16\pi}g^{\alpha\mu}g^{\beta\nu}f_{\alpha\beta}f_{\mu\nu}. \qquad (6.61)$$

6.1. ELEMENTS OF GENERAL RELATIVITY

It can easily be shown to be given by

$$T_{\rho\sigma} = \frac{1}{4\pi} \left(\frac{1}{4} g_{\rho\sigma} f_{\alpha\beta} f^{\alpha\beta} - f_{\rho\alpha} f_\sigma{}^\alpha \right). \tag{6.62}$$

If we calculate the trace of the energy-momentum tensor (6.62) we find that it vanishes,

$$T = T_\rho{}^\rho = g^{\rho\sigma} T_{\rho\sigma} = 0. \tag{6.63}$$

Using now $R = -\kappa T$ then leads to the vanishing of the Ricci scalar curvature, $R = 0$. We therefore obtain

$$R_{\mu\nu} = \kappa T_{\mu\nu} \tag{6.64}$$

for the Einstein field equations in the presence of an electromagnetic field. In Eq. (6.64) the energy-momentum tensor $T_{\mu\nu}$ is given by Eq. (6.62).

The Einstein field equations (6.64) and the Maxwell equations (5.28) (Chapter 5) constitute *the coupled Einstein-Maxwell field equations*.

6.1.5 The Schwarzschild Solution

In spite of the nonlinearity of the Einstein field equations, there are numerous exact solutions to these equations. Moreover, there are other solutions which are not exact but approximate. Exact solutions are usually obtained using special methods.

The simplest of all exact solutions to Einstein's field equations is that of Schwartzschild. The solution is *spherically symmetric* and *static*. Such a field can be produced by a spherically symmetric distribution and motion of matter. It follows that the requirement of spherical symmetry alone is sufficient to yield a static solution.

The spherical symmetry of the metric means that the expression for the interval $ds = (g_{\mu\nu} dx^\mu dx^\nu)^{1/2}$ must be the same for all points located at the same distance from the center. In flat space their distance is equal to the radius vector, and the metric is given by (c is taken as equal to 1):

$$ds^2 = dt^2 - dr^2 - r^2 \left(d\theta^2 + \sin^2\theta d\phi^2 \right). \tag{6.65}$$

In a non-Euclidean space, such as the Riemannian one we have in the presence of a gravitational field, there is no quantity which has all the properties of the flat space radius vector, such as that it is equal both to the distance from the center and to the length of the circumference divided by 2π. Therefore, the choice of a radius vector is here arbitrary.

When a mass with spherical symmetry is introduced at the origin, the flat space line element (6.65) must be modified but in a way that retains spherical symmetry. The most general spherically symmetric expression for ds^2 is

$$ds^2 = a(r,t)\,dt^2 + b(r,t)\,dr^2 + c(r,t)\,drdt + d(r,t)\left(d\theta^2 + \sin^2\theta d\phi^2\right). \tag{6.66}$$

Because of the arbitrariness in the choice of the coordinate system in general relativity theory, we can perform a coordinate transformation which does not destroy the spherical symmetry of ds^2. Hence we can choose new coordinates r' and t' given by some functions $r' = r'(r,t)$ and $t' = t'(r,t)$.

Making use of these transformations, we can choose the new coordinates so that the coefficient $c(r,t)$ of the mixed term $drdt$ vanishes and the coefficient $d(r,t)$ of the angular part to be $-r'^2$, in the metric (6.66). The latter condition implies that the radius vector is now defined in such a way that the circumference of a circle whose center is at the origin of the coordinates is equal to $2\pi r$. It is convenient to express the functions $a(r,t)$ and $b(r,t)$ in exponential forms, e^ν and $-e^\lambda$, respectively, where ν and λ are functions of the new coordinates r' and t'. Consequently, the line element (6.66) will have the form

$$ds^2 = e^\nu dt^2 - e^\lambda dr^2 - r^2\left(d\theta^2 + \sin^2\theta d\phi^2\right), \tag{6.67}$$

where, for brevity, we have dropped the primes from the new coordinates r' and t', and the speed of light c is taken as equal to 1.

We now denote the coordinates t, r, θ, ϕ by x^0, x^1, x^2, x^3, respectively. Hence the components of the covariant metric tensor are given by:

$$g_{\mu\nu} = \begin{pmatrix} e^\nu & 0 & 0 & 0 \\ 0 & -e^\lambda & 0 & 0 \\ 0 & 0 & -r^2 & 0 \\ 0 & 0 & 0 & -r^2\sin^2\theta \end{pmatrix}, \tag{6.68a}$$

whereas those of the contravariant metric tensor are:

$$g^{\mu\nu} = \begin{pmatrix} e^{-\nu} & 0 & 0 & 0 \\ 0 & -e^{-\lambda} & 0 & 0 \\ 0 & 0 & -r^{-2} & 0 \\ 0 & 0 & 0 & -r^{-2}\sin^{-2}\theta \end{pmatrix}. \tag{6.68b}$$

To find out the differential equations that the functions ν and λ have to satisfy, according to Einstein's field equations, we first need to calculate

6.1. ELEMENTS OF GENERAL RELATIVITY

the Christoffel symbols associated with the metric (6.68). The nonvanishing components are:

$$\Gamma^0_{00} = \frac{\dot\nu}{2}, \quad \Gamma^0_{10} = \frac{\nu'}{2}, \quad \Gamma^0_{11} = \frac{\dot\lambda}{2}e^{\lambda-\nu}, \tag{6.69a}$$

$$\Gamma^1_{00} = \frac{\nu'}{2}e^{\nu-\lambda}, \quad \Gamma^1_{10} = \frac{\dot\lambda}{2}, \quad \Gamma^1_{11} = \frac{\lambda'}{2}, \tag{6.69b}$$

$$\Gamma^1_{22} = -re^{-\lambda}, \quad \Gamma^1_{33} = -r\sin^2\theta e^{-\lambda}, \quad \Gamma^2_{12} = \frac{1}{r}, \tag{6.69c}$$

$$\Gamma^2_{33} = -\sin\theta\cos\theta, \quad \Gamma^3_{13} = \frac{1}{r}, \quad \Gamma^3_{23} = \cot\theta, \tag{6.69d}$$

where dots and primes denote differentiation with respect to t and r, respectively.

With these Christoffel symbols, we compute the following expressions for the nonvanishing components of the Einstein tensor:

$$G_0{}^0 = -e^{-\lambda}\left(\frac{1}{r^2} - \frac{\lambda'}{r}\right) + \frac{1}{r^2} = \kappa T_0{}^0, \tag{6.70a}$$

$$G_0{}^1 = -e^{-\lambda}\frac{\dot\lambda}{r} = \kappa T_0{}^1, \tag{6.70b}$$

$$G_1{}^1 = -e^{-\lambda}\left(\frac{\nu'}{r} + \frac{1}{r^2}\right) + \frac{1}{r^2} = \kappa T_1{}^1, \tag{6.70c}$$

$$G_2{}^2 = -\frac{1}{2}e^{-\lambda}\left(\nu'' + \frac{\nu'^2}{2} + \frac{\nu'-\lambda'}{r} - \frac{\nu'\lambda'}{2}\right)$$
$$+\frac{1}{2}e^{-\nu}\left(\ddot\lambda + \frac{\dot\lambda^2}{2} - \frac{\dot\lambda\dot\nu}{2}\right) = \kappa T_2{}^2, \tag{6.70d}$$

$$G_3{}^3 = G_2{}^2 = \kappa T_3{}^3. \tag{6.70e}$$

All other components vanish identically.

The gravitational field equations can now be integrated exactly for the spherical symmetric field in vacuum, i.e., outside the masses producing the field. Setting Eqs. (6.70) equal to zero leads to the independent equations:

$$e^{-\lambda}\left(\frac{\nu'}{r} + \frac{1}{r^2}\right) - \frac{1}{r^2} = 0, \tag{6.71a}$$

$$e^{-\lambda}\left(\frac{\lambda'}{r} - \frac{1}{r^2}\right) + \frac{1}{r^2} = 0, \qquad (6.71b)$$

$$\dot{\lambda} = 0. \qquad (6.71c)$$

From Eq. (6.71a) and (6.71b) we find $\nu' + \lambda' = 0$, so that $\nu + \lambda = f(t)$, where $f(t)$ is a function of t only. If we perform now the coordinate transformation $x^0 = h(x'^0)$, $x^k = x'^k$, then $g'_{00} = \dot{h}^2 g_{00}$. Such a transformation amounts to adding to the function ν an arbitrary function of time, while leaving unaffected the other components of the metric. Hence we can choose the function h so that $\nu + \lambda = 0$. Consequently, we see, by Eq. (6.71c), that both ν and λ are time-independent. In other words the spherically symmetric gravitational field in vacuum is automatically static.

Equation (6.71b) can now be integrated. It gives:

$$e^{-\lambda} = e^{\nu} = 1 - \frac{K}{r}, \qquad (6.72)$$

where K is an integration constant. We see that for $r \to \infty$, $e^{-\lambda} = e^{\nu} = 1$, i.e., far from the gravitational bodies, the metric reduces to that of the flat space (6.65). The constant K can easily be determined from the requirement that Newton's law of motion be obtained at large distances from the central mass. From the geodesic equation it follows that the radial acceleration of a small test mass at rest with respect to the central mass is (see Problem 6.20):

$$-\Gamma^1_{00} = -\frac{1}{2}\left(1 - \frac{K}{r}\right)\frac{K}{r^2} \to -\frac{K}{2r^2}. \qquad (6.73)$$

Comparing this expression with the Newtonian value $-Gm/r^2$ gives $K = 2Gm$, where m is the central mass and G is the Newton constant.

The constant $2Gm$, or $2Gm/c^2$ in units where c is not taken as equal to 1, is often called the *Schwarzschild radius* of the mass m. For example, the Schwarzschild radius for the Sun is 2.95 km, that for the Earth is 8.9 mm, and that for an electron is 13.5×10^{-56} cm.

We therefore obtain for the spherically symmetric metric the form:

$$g_{\mu\nu} = \begin{pmatrix} 1 - 2Gm/r & 0 & 0 & 0 \\ 0 & -(1 - 2Gm/r)^{-1} & 0 & 0 \\ 0 & 0 & -r^2 & 0 \\ 0 & 0 & 0 & -r^2 \sin^2\theta \end{pmatrix}. \qquad (6.74)$$

It is known as the *Schwarzschild solution* and describes the most general spherically symmetric solution of the Einstein field equations in a region

6.1. ELEMENTS OF GENERAL RELATIVITY

of space where the energy-momentum tensor $T^{\mu\nu}$ vanishes. Although $g_{\mu\nu}$ goes to the flat space metric when r goes to infinity, it was *not* necessary to require this asymptotic behavior to obtain the solution.

It is worth mentioning that all spherically symmetric solutions of the Einstein field equations in vacuum which satisfy the boundary conditions at infinity mentioned above are equivalent to the Schwarzschild field, i.e., their time-dependence can be eliminated by a suitable coordinate transformation. This result is due to Birkhoff.

Finally, it is convenient to introduce Cartesian coordinates by means of the coordinate transformation

$$\begin{aligned} x^1 &= r\sin\theta\cos\phi, \\ x^2 &= r\sin\theta\sin\phi, \\ x^3 &= r\cos\theta. \end{aligned} \qquad (6.75)$$

In terms of these coordinates, the Schwarzschild metric (6.74) will then have the form

$$g_{00} = 1 - \frac{2Gm}{r},$$

$$g_{0r} = 0, \qquad (6.76)$$

$$g_{rs} = -\delta_{rs} - \frac{2Gm/r}{1-2Gm/r}\frac{x^r x^s}{r^2}.$$

6.1.6 Experimental Tests of General Relativity

Up to a few years ago, general relativity was verified by three tests: the gravitational red shift, the deflection of light near massive bodies and the planetary orbit effect on the planets. The first could also be explained, in fact, without the use of the Einstein field equations. However, this picture has been changed.

Gravitational Red Shift

Consider the clocks at rest at two points 1 and 2. The rate of change of times at these points are then given by $ds(1) = \sqrt{g_{00}(1)}dt$ and $ds(2) = \sqrt{g_{00}(2)}dt$. The relation between the rates of identical clocks in a gravitational field is therefore given by $\sqrt{g_{00}(2)/g_{00}(1)}$. The frequency of an atom, ν_0, located at point 1, when seen by an observer at point 2 is, hence,

given by
$$\nu = \nu_0 \sqrt{\frac{g_{00}(1)}{g_{00}(2)}}. \tag{6.77}$$

For a gravitational field like that of Schwarzschild, one therefore obtains for the frequency shift per unit frequency:

$$\frac{\Delta\nu}{\nu_0} = \frac{\nu - \nu_0}{\nu_0} \approx -\frac{Gm}{c^2}\left(\frac{1}{r_1} - \frac{1}{r_2}\right), \tag{6.78}$$

to first order in $Gm/c^2 r$. If we take r_1 to be the observed radius of the Sun and r_2 the radius of the Earth's orbit around the Sun (thus neglecting completely the Earth's gravitational field), then $\Delta\nu/\nu_0 = -2.12 \times 10^{-6}$. This frequency shift is usually referred to as the *gravitational red shift*.

The gravitational red shift was tested for the Sun and for white dwarfs, and it was suggested that it be tested by atomic clocks. The red shift was also observed directly using the Mössbauer effect by Pound and Rabka, and by Cranshaw, Schiffer and Whitehead. The latter employed Fe^{57} and a total height difference of 12.5 metres. A red shift 0.96 ± 0.45 times the predicted value was observed by them. Pound and Rabka's result is more precise. They obtained a red shift 1.05 ± 0.10 times the predicted value.

Effects on Planetary Motion

One assumes that test particles move along geodesics in the gravitational field (see next subsection), and that planets have small masses as compared with the mass of the Sun, thus behaving like test particles. Consequently, to find the equation of motion of a planet moving in the gravitational field of the Sun one has to write the geodesic equation in the Schwarzschild field. In fact one does not need the exact solution (6.76) but its first approximation,

$$\begin{aligned} g_{00} &= 1 - 2Gm/r, \\ g_{0r} &= 0, \\ g_{rs} &= -\delta_{rs} - 2Gm x^r x^s/r^3. \end{aligned} \tag{6.79}$$

In the above equations the speed of light is taken as unity.

Using the approximate metric (6.79) in the geodesic equation (6.43) gives (see Problem 6.24)

$$\ddot{\mathbf{x}} - Gm\nabla\frac{1}{r}$$
$$= Gm\left\{2(\dot{\mathbf{x}}^2)\nabla\frac{1}{r} - 2Gm\frac{1}{r}\nabla\frac{1}{r} - 2\left(\dot{\mathbf{x}}\cdot\nabla\frac{1}{r}\right)\dot{\mathbf{x}} + \frac{3}{r^5}(\mathbf{x}\cdot\dot{\mathbf{x}})^2\mathbf{x}\right\}, \tag{6.80}$$

6.1. ELEMENTS OF GENERAL RELATIVITY

where we have used three-dimensional notation and a dot denotes differentiation with respect to t. Multiplying Eq. (6.80), vectorially, by the radius vector $\mathbf{x}$ gives

$$\mathbf{x} \times \ddot{\mathbf{x}} = -2Gm \left(\dot{\mathbf{x}} \cdot \nabla \frac{1}{r} \right) (\mathbf{x} \times \dot{\mathbf{x}}), \tag{6.81}$$

thus leading to the first integral

$$\mathbf{x} \times \dot{\mathbf{x}} = \mathbf{J} e^{-2Gm/r}, \tag{6.82}$$

where $\mathbf{J}$ is a *constant* vector, the angular momentum per mass unit.

Hence the radius vector $\mathbf{x}$ moves in a plane perpendicular to the vector $\mathbf{J}$, as in Newtonian mechanics. Introducing in this plane polar coordinates r, ϕ to describe the motion of the planet, the equation of motion (6.80), consequently, decomposes into

$$\ddot{r} - r\dot{\phi}^2 + \frac{Gm}{r^2} = \frac{Gm}{r^2} \left\{ 3\dot{r}^2 - 2r^2\dot{\phi}^2 + 2\frac{Gm}{r} \right\}, \tag{6.83a}$$

$$r^2 \dot{\phi} = J e^{-2Gm/r}, \tag{6.83b}$$

where J is the magnitude of the vector $\mathbf{J}$.

Introducing now the new variable $u = 1/r$ one can rewrite Eqs. (6.83) in terms of $u(\phi)$:

$$u'' + u - \frac{Gm}{J^2} = Gm \left(-u'^2 + 2u^2 + 2\frac{Gm}{J^2} u \right). \tag{6.84}$$

Here a prime denotes a derivation with respect to the angle ϕ.

Let us try a solution of the form

$$u = b(1 + \epsilon \cos \alpha \phi). \tag{6.85}$$

Here ϵ is the eccentricity and α is some parameter to be determined and whose value in the usual nonrelativistic mechanics is unity. The other constant b is related to J in the nonrelativistic mechanics by $Gm/J^2 = b$. Using the above solution in Eq. (6.84) and equating coefficients of $\cos \alpha \phi$ gives

$$\alpha^2 = 1 - 2Gm \left(2b + Gm/J^2 \right). \tag{6.86}$$

Substituting for Gm/J^2 its nonrelativistic value b then gives $\alpha^2 = 1 - 6Gmb$, or, to a first approximation in Gm,

$$\alpha = 1 - 3Gmb. \tag{6.87}$$

Successive perihelia occur when

$$(1 - 3Gmb)(2\pi + \Delta\phi) = 2\pi. \tag{6.88}$$

Consequently, there will be an advance in the perihelion of the orbit per revolution given by $\Delta\phi = 6\pi Gmb$, or $\Delta\phi = 6\pi Gm/a\,(1 - \epsilon^2)$ if we make use of the nonrelativistic value of the constant b, and where a is the semimajor axis of the orbit. Reinstating now c, the velocity of light, finally gives for the perihelion advance

$$\Delta\phi = \frac{6\pi Gm}{c^2 a\,(1 - \epsilon^2)}, \tag{6.89}$$

in radians per revolution.

We list below the calculated values of $\Delta\phi$ per century for four planets:

Planet	$\Delta\phi$
Mercury	43.03″
Venus	8.60″
Earth	3.80″
Mars	1.35″

The astronomical observations for the planet Mercury give 43.11 ± 0.45 sec per century, in good agreement with the calculated value.

Deflection of Light

To discuss the deflection of light in the gravitational field we must again solve the geodesic equation, but now with the null conditions $ds = 0$. Using the appropriate solution (6.79) then gives for $g_{\mu\nu}dx^\mu dx^\nu = 0$

$$\left(1 + \frac{2Gm}{r}\right)\left[\dot{\mathbf{x}}\cdot\dot{\mathbf{x}} + \frac{2Gm}{r^3}(\mathbf{x}\cdot\dot{\mathbf{x}})^2\right] = 1. \tag{6.90}$$

Using polar coordinates r, ϕ, consequently, gives to the first approximation in Gm

$$\dot{r}^2 + r^2\dot{\phi}^2 + \frac{4Gm\dot{r}^2}{r} + 2Gmr\dot{\phi}^2 = 1. \tag{6.91}$$

Again changing variables into $u(\phi) = 1/r$, and using Eq. (6.83b), gives

$$u'^2 + u^2 + 2Gmu\left(2u'^2 + u^2\right) = J^{-2}e^{4Gmu}. \tag{6.92}$$

6.1. ELEMENTS OF GENERAL RELATIVITY

Differentiation of this equation with respect to ϕ gives

$$u'' + u + Gm\left(2u'^2 + 4uu'' + 3u^2\right) = 2GmJ^{-2}, \qquad (6.93)$$

to the first approximation in Gm.

To solve Eq. (6.93) we note that in the lowest approximation one has

$$u'^2 \approx J^{-2} - u^2, \qquad (6.94)$$

$$u'' \approx -u. \qquad (6.95)$$

Using these values in Eq. (6.93) gives

$$u'' + u = 3Gmu^2, \qquad (6.96)$$

for the orbit of the light ray. In the lowest approximation u satisfies $u''+u = 0$, whose solution is a straight line

$$u = \frac{\cos\phi}{R}, \qquad (6.97)$$

where R is a constant. This shows that $r = 1/u$ has a minimum value R at $\phi = 0$. Substituting into the right-hand side of Eq. (6.96) then gives

$$u'' + u = 3\frac{Gm}{R^2}\cos^2\phi. \qquad (6.98)$$

The solution of this equation is

$$u = \frac{\cos\phi}{R} + \frac{Gm}{R^2}\left(1 + \sin^2\phi\right). \qquad (6.99)$$

Introducing now Cartesian coordinates $x = r\cos\phi$ and $y = r\sin\phi$, the above equation gives

$$x = R - \frac{Gm}{R}\frac{x^2 + 2y^2}{\sqrt{x^2 + y^2}}. \qquad (6.100)$$

For large values of $|y|$ this equation becomes

$$x \approx R - \frac{2Gm}{R}|y|. \qquad (6.101)$$

Hence, asymptotically, the orbit of the light ray is a straight line in space. This result is expected, since far away from the central mass the space is flat. The angle $\Delta\phi$ between the two asymptotes is, however, equal to

$$\Delta\phi = 4\frac{Gm}{c^2 R}, \qquad (6.102)$$

in units in which c is different from unity.

The angle $\Delta\phi$ represents the angle of *deflection* of a light ray in passing through the Schwarzschild field. For a light ray just grazing the Sun Eq. (6.102) gives $\Delta\phi = 1.75$ sec. Observations indeed confirm this result; one of the latest results gives 1.75 ± 0.10 sec.

Gravitational Radiation Experiments

Weber has developed methods to detect gravitational waves that Einstein's gravitational field equations predict. The experiment involves detectors at opposite ends of a 1000 km baseline. Sudden increases in detector output were observed by him roughly once in several days, coincident within the resolution time of 0.25 seconds.

Weber's apparatus measures the Fourier transform of the Riemann tensor. The method uses the fact that the distance η^μ between two neighbouring test particles, which follow geodesics, satisfies the *geodesic deviation* equation

$$\frac{\delta^2 \eta^\mu}{\delta s^2} + R^\mu{}_{\alpha\nu\beta}\lambda^\alpha \eta^\nu \lambda^\beta = 0, \qquad (6.103)$$

where λ^α is the tangent vector to one of the geodesics, and $\delta/\delta s = \lambda^\alpha \nabla_\alpha$ is a directional covariant derivative. Weber measured the strains of a large aluminium cylinder, having mass of the order 10^6 grams, by means of a piezoelectric crystal attached to the cylinder which transforms the mechanical movement into an electric current. The detector was developed for operation in the vicinity of 1662 cycle/sec. A high frequency source was developed for dynamic gravitational fields and the detector was tested by doing a communication experiment with high frequency Coulomb fields.

Radar Experiment

Shapiro has designed a radar experiment to test general relativity by measuring the effect of solar gravity on time delays of round-trip travel times of radar pulses transmitted from the Earth toward an inner planet, i.e., Venus or Mercury. The experiment is based on the phenomenon that electromagnetic waves "slow down" in the gravitational field. Within the framework of general relativity there should be an anomalous delay of 200 microseconds

6.1. ELEMENTS OF GENERAL RELATIVITY

in the arrival time of a radar echo from Mercury, positioned on the far side of the Sun near the limb.

For example, if we calculate the proper time τ at $r = r_2$ for a radial round-trip travel $r_2 \to r_1 \to r_2$, with $r_2 > r_1$, of a radar pulse in the Schwarzschild field, and subtract from τ the corresponding value τ_0 when the spherical mass $m = 0$, we find

$$\Delta \tau = \frac{4Gm}{c^3} \left(\ln \frac{r_2}{r_1} - \frac{r_2 - r_1}{r_2} \right) + O\left(m^2\right). \qquad (6.104)$$

In general one finds

$$\Delta \tau \approx \frac{4Gm}{c^3} \left(\ln \frac{r_e + r_p + R}{r_e + r_p - R} \right), \qquad (6.105)$$

where r_e is the Earth-Sun distance, r_p the planet-Sun distance, and R the Earth-planet distance.

Shapiro found that the retardation of radar signals are 1.02 ± 0.05 times the corresponding effect predicted by general relativity.

Low-Temperature Experiments

Schiff has proposed an experiment to check the equations of motion in general relativity by means of a gyroscope, which is forced to go around the Earth either in a stationary laboratory fixed to the Earth or a satellite. The unique experiment is made possible by complete use of a low-temperature environment, and the properties of superconductors, including the use of zero magnetic fields and ultrasensitive magnetometry. Schiff has calculated, using results obtained by Papapetrou for the motion of spinning bodies in general relativity, that a perfect gyroscope subject to no torques will experience an anomalous precession with respect to the fixed stars as it travels around the Earth.

6.1.7 Equations of Motion

Geodesic Postulate

In the last subsection it was assumed that the planet's motion around the Sun is described by the geodesic equation (6.43). The assumption that the equations of motion of a test particle, moving in gravitational field, are given by the geodesic equation is known as the *geodesic postulate* and was suggested by Einstein in his first article on the general theory of relativity.

Eleven years later when Einstein and Grommer showed that the geodesic postulate need not be assumed, but that it rather follows from the gravitational field equations; this is a consequence of the nonlinearity of the field equations along with the fact that they satisfy the four contracted Bianchi identities (see Subsection 6.1.1). The discovery of Einstein and Grommer is considered to be one of the most important achievements, and one of the most attractive features of the general theory of relativity. Later on Infeld and Schild showed that the equations of motion of a test particle are given by the geodesic equation in the *external gravitational* field. This result, however, does not differ from the geodesic postulate because, by definition, a test particle has no self-field.

Equations of Motion as a Consequence of Field Equations

In order to establish the relation between the Einstein field equations and the equations of motion one proceeds as follows. We have seen in Subsection 6.1.4 that because of the contracted Bianchi identity it follows that the energy-momentum tensor $T^{\mu\nu}$ satisfies a generally covariant conservation law of the form given by Eq. (6.51). Consequently, one obtains for the energy-momentum tensor *density* $\mathcal{T}^{\mu\nu}$

$$\nabla_\nu \mathcal{T}^{\mu\nu} = \partial_\nu \mathcal{T}^{\mu\nu} + \Gamma^\mu_{\alpha\beta} \mathcal{T}^{\alpha\beta} = 0, \qquad (6.106)$$

where $\mathcal{T}^{\mu\nu} \equiv \sqrt{-g} T^{\mu\nu}$.

For a system of N particles of finite masses, represented as singularities of the gravitational field, $\mathcal{T}^{\mu\nu}$ may be taken in the form

$$\mathcal{T}^{\mu\nu} = \sum_{A=1}^{N} m_A v_A^\mu v_A^\nu \delta_A (\mathbf{x} - \mathbf{z}_A). \qquad (6.107)$$

Here z_A^μ are the coordinates of the Ath particle. (Roman capital indices, $A, B, \cdots$, run from 1 to N. For these indices the summation convention will be suspended.) Also $v^\mu = \dot{z}^\mu = dz^\mu/dt$ ($v_A^0 = \dot{z}_A^0 = 1$), and δ is the three-dimensional Dirac delta function satisfying the following conditions:

$$\delta(\mathbf{x}) = 0; \quad \text{for} \quad x \neq 0, \qquad (6.108a)$$

$$\int \delta(\mathbf{x} - \mathbf{z}) d^3x = 1, \qquad (6.108b)$$

$$\int f(\mathbf{x}) \delta(\mathbf{x} - \mathbf{z}) d^3x = f(\mathbf{z}), \qquad (6.108c)$$

6.1. ELEMENTS OF GENERAL RELATIVITY

for any continuous function $f(\mathbf{x})$ in the neighbourhood of $\mathbf{z}$. In Eq. (6.107), m_A is a function of time which may be called the *inertial mass* of the Ath particle.

If we put the energy-momentum tensor density (6.107) into (6.106) and integrate over the three-dimensional region surrounding the first singularity, we obtain

$$\frac{dp^\mu}{dt} = \int F^\mu \delta(\mathbf{x} - \mathbf{z}) \, d^3x, \qquad (6.109)$$

where $p^\mu = mv^\mu$ and $F^\mu = -m\Gamma^\mu_{\alpha\beta}v^\alpha v^\beta$, and where we have put, for simplicity, $m = m_1$, $z^\mu = z_1^\mu$, $v^\mu = v_1^\mu$, and $\delta(\mathbf{x} - \mathbf{z}) = \delta_1(\mathbf{x} - \mathbf{z}_1)$.

Self-Action Terms

Equation (6.109) may be interpreted as an "exact equation of motion" of the first particle. However, since the Christoffel symbols are singular at the location of the particle, the equation contains infinite self-action terms. However, it was shown by Carmeli that these terms can be removed as follows.

Putting Eq. (6.107) into Eq. (6.106) we obtain

$$\partial_0 \left[\sum_{A=1}^N m_A v_A^\mu \delta_A \right] + \partial_n \left[\sum_{A=1}^N m_A v_A^\mu v_A^n \delta_A \right] + \sum_{A=1}^N m_A \Gamma^\mu_{\alpha\beta} v_A^\alpha v_A^\beta \delta_A = 0, \qquad (6.110)$$

where Latin indices run from 1 to 3. The first term on the left-hand side of Eq. (6.106) can be written as

$$\partial_0 \left[\sum_{A=1}^N m_A v_A^\mu \delta_A \right] = \sum_{A=1}^N \partial_0 (m_A v_A^\mu) \delta_A + \sum_{A=1}^N m_A v_A^\mu \partial_0 \delta_A, \qquad (6.111)$$

with

$$\partial_0 \delta_A = \partial_0 \delta_A (x^s - z_A^s) = -\partial_n \delta_A v_A^n. \qquad (6.112)$$

Using the above results in Eq. (6.110), we obtain

$$\sum_{A=1}^N \left\{ \frac{d(m_A v_A^\mu)}{dt} + m_A \Gamma^\mu_{\alpha\beta} v_A^\alpha v_A^\beta \right\} \delta_A = 0. \qquad (6.113)$$

Equation (6.113), which is identical with Eq. (6.106), is satisfied for any spacetime point, since otherwise the Bianchi identity or the Einstein field equations will not be satisfied.

We now examine the behavior of Eq. (6.113) in the infinitesimal neighbourhood of the first singularity, which we assume not to contain any other singularity. In this region $\delta_B(\mathbf{x} - \mathbf{z}_B) = 0$ for $B = 2, 3, \cdots, N$. Hence Eq. (6.113) gives for the conservation law near the first singularity

$$\left\{ \frac{d(mv^\mu)}{dt} + m\Gamma^\mu_{\alpha\beta} v^\alpha v^\beta \right\} \delta(\mathbf{x} - \mathbf{z}) = 0. \tag{6.114}$$

Let us further assume that the Christoffel symbols near the first singularity can be expanded into a power series in the infinitesimal distance r, defined by $r^2 = (x^s - z^s)(x^s - z^s)$, where $z^s = z_1^s$, in the vicinity of the first particle. Then we have

$$\Gamma^\mu_{\alpha\beta} = {}_{-k}\Gamma^\mu_{\alpha\beta} + {}_{-k+1}\Gamma^\mu_{\alpha\beta} + \cdots + {}_0\Gamma^\mu_{\alpha\beta} + \cdots, \tag{6.115}$$

where the indices written in subscripts on the left of a function indicate its behavior with respect to r, and k is a positive integer.

For example ${}_0\Gamma^\mu_{\alpha\beta}$ is that part of the Christoffel symbol which varies as r^0, i.e., is finite at the location of the first particle. When one uses spherical coordinates r, θ and ϕ, one can write

$${}_{-k}\Gamma^\mu_{\alpha\beta} = r^{-k} A^\mu_{\alpha\beta}(\theta, \phi), \tag{6.116a}$$

$${}_{-k+1}\Gamma^\mu_{\alpha\beta} = r^{-k+1} B^\mu_{\alpha\beta}(\theta, \phi), \tag{6.116b}$$

$$\cdot \quad \cdot \quad \cdot \quad \cdot \quad \cdot \quad \cdot$$

$${}_0\Gamma^\mu_{\alpha\beta} = D^\mu_{\alpha\beta}(\theta, \phi), \text{ etc.} \tag{6.116c}$$

Terms like ${}_1\Gamma^\mu_{\alpha\beta}$, ${}_2\Gamma^\mu_{\alpha\beta}$, etc., however, need not be taken into account when one puts the above expansion into Eq. (6.114) since $r^j \delta(\mathbf{x} - \mathbf{z}) = 0$ for any positive integer j. If we denote now $mA^\mu_{\alpha\beta} v^\alpha v^\beta, \cdots$ by $A^\mu, \cdots$ we can write Eq. (6.114) in the form

$$\left\{ r^{-k} A^\mu + r^{-k+1} B^\mu + \cdots + r^{-1} C^\mu + D_1^\mu \right\} \delta(\mathbf{x} - \mathbf{z}) = 0, \tag{6.117}$$

where we have used the notation $D_1^\mu = d(mv^\mu)/dt + D^\mu$.

In order to get rid of terms proportional to negative powers of r in Eq. (6.117) we proceed as follows. Multiplying Eq. (6.117) by r^k and using $r^j \delta(\mathbf{x} - \mathbf{z}) = 0$ we obtain

$$A^\mu(\theta, \phi) \delta(\mathbf{r}) = 0, \tag{6.118}$$

6.1. ELEMENTS OF GENERAL RELATIVITY

the integration of which over the three-dimensional region yields, using spherical coordinates,

$$\int\int A^\mu(\theta,\phi)\sin\theta d\theta d\phi \int r^2 \delta(\mathbf{r})=0. \tag{6.119}$$

From the property of the delta-function

$$\int \delta(\mathbf{r})d^3x = \int\int \sin\theta d\theta d\phi \int \delta(\mathbf{r})r^2 dr = 1, \tag{6.120}$$

one obtains $\int \delta(\mathbf{r})r^2 dr = (4\pi)^{-1}$. Hence we obtain

$$\int\int A^\mu(\theta,\phi)\sin\theta d\theta d\phi = 0, \tag{6.121}$$

independent of the value of the variable R. Thus the angular distribution of $A^\mu(\theta,\phi)$ is such that its average equals zero.

However, not only does the above equation hold, but also (s is any finite positive integer)

$$a(r) = r^{-s}\int\int A^\mu(\theta,\phi)\sin\theta d\theta d\phi = 0, \tag{6.122}$$

for small values of r as well as when r tends to zero, as can be verified by using L'Hospital's theorem, for example. It follows then that $a(r)$ is a function of r whose value is zero for any small r, including $r=0$. Using the property of delta-function we obtain

$$\int r^2 \delta(\mathbf{r})dr = (4\pi)^{-1} f(0), \tag{6.123}$$

for any continuous function of r. Since $a(r)$ is certainly continuous, one obtains

$$\int r^2\delta(\mathbf{r})a(r)dr = 0. \tag{6.124}$$

Hence when one integrates Eq. (6.117) over the three-dimensional space, there will be no contribution from the first term.

In order to show that the second term of Eq. (6.117) will also not contribute to the three-dimensional integration of the same equation, we multiply it by r^{k-1}. We obtain now, after neglecting terms that do not contribute,

$$\{r^{-1}A^\mu(\theta,\phi) + B^\mu(\theta,\phi)\}\delta(\mathbf{r}) = 0. \tag{6.125}$$

Integration of this equation, again using spherical coordinates, shows that the first term will not contribute anything because of Eq. (6.124), and we are left with

$$\int\int B^\mu(\theta,\phi)\sin\theta d\theta d\phi \int r^2\delta(\mathbf{r})\,dr = 0. \tag{6.126}$$

Hence we have

$$\int\int B^\mu(\theta,\phi)\sin\theta d\theta d\phi = 0, \tag{6.127}$$

independent of r. From this equation one obtains another one, analogous to Eq. (6.124) but with B^μ instead of A^μ:

$$\int r^2\delta(\mathbf{r})b(r)\,dr = 0, \tag{6.128}$$

with

$$b(r) = r^{-s}\int\int B^\mu(\theta,\phi)\sin\theta d\theta d\phi = 0. \tag{6.129}$$

Proceeding in this way, one verifies that the angular distribution of all functions A^μ, B^μ, etc., is such that they all satisfy equations like Eqs. (6.121) and (6.127). Hence it is clear that one obtains

$$\int D_1^\mu(\theta,\phi)\delta(\mathbf{r})\,d^3x = 0, \tag{6.130}$$

which gives

$$\frac{dp^\mu}{dt} + mv^\alpha v^\beta \int {}_0\Gamma^\mu_{\alpha\beta}\delta(\mathbf{r})\,d^3x = 0, \tag{6.131}$$

or equivalently

$$\dot{v}^k + v^\alpha v^\beta \int \left({}_0\Gamma^k_{\alpha\beta} - v^k{}_0\Gamma^0_{\alpha\beta}\right)\delta(\mathbf{r})\,d^3x = 0. \tag{6.132}$$

Equation (6.132) is the "exact equation of motion".

Einstein-Infeld-Hoffmann Method

Having found the law of motion (6.131), one can now proceed to find the equation of motion of two finite masses, each moving in the field produced by both of them. In the following we find such an equation of motion in the case for which the particles' velocities are much smaller than the speed of light. Moreover, we will confine ourselves to an accuracy of post-Newtonian. This means the equation of motion obtained will contain the Newtonian

6.1. ELEMENTS OF GENERAL RELATIVITY

equation as a limit, but is a first generalization of it. Such an equation was first obtained by Einstein, Infeld, and Hoffmann. To obtain this equation we solve the field equations and formulate the equations of motion explicitly by means of an approximation method, the Einstein-Infeld-Hoffmann (EIH) method, to be described below.

Let us assume a function ϕ developed in a power series in the parameter $\lambda = 1/c$, where c is the speed of light. One then has

$$\phi = {}_0\phi + {}_1\phi + {}_2\phi + \cdots. \tag{6.133}$$

The indices written as left subscripts indicate the order of λ absorbed by the ϕ's.

If a function $\phi(x)$ varies rapidly in space but slowly with x^0, then we are justified in not treating all its derivatives in the same manner. The derivatives with respect to x^0 will be of a higher order than the space derivatives. We thus write

$$\partial_0\left({}_l\psi\right) = {}_{l+1}\psi. \tag{6.134}$$

That is, differentiation with respect to x^0 raises the order by one. Thus if the coordinates z^s of a particle are considered to be of order zero, $\dot{z}^s$ will be of order one, and $\ddot{z}^s$ of order two. Using now the Newtonian approximation mass $\times$ acceleration=mass $\times$ mass/(distance)^2, we see the mass is of order two. In all the power developments we take into account only even or only odd powers of $1/c$. (The expansion of the metric tensor, etc., in a power series in c^{-2} (such as $\phi = {}_0\phi + {}_2\phi + \cdots$, or $\phi = {}_1\phi + {}_3\phi + \cdots$) corresponds to the choice of the symmetric Green function, thus excluding radiation.)

Thus, because of the order with which we start m and $\dot{z}^s$, we have

$$T^{00} = {}_2T^{00} + {}_4T^{00} + \cdots,$$
$$T^{0n} = {}_3T^{0n} + {}_5T^{0n} + \cdots, \tag{6.135}$$
$$T^{mn} = {}_4T^{mn} + {}_6T^{mn} + \cdots.$$

As to the metric tensor, we write

$$g_{\mu\nu} = \eta_{\mu\nu} + h_{\mu\nu}, \quad g^{\mu\nu} = \eta^{\mu\nu} + h^{\mu\nu}. \tag{6.136}$$

The gravitational field equations can be written as

$$\sqrt{-g}R_{\alpha\beta} = \kappa\left(T_{\alpha\beta} - \frac{1}{2}g_{\alpha\beta}T\right), \tag{6.137}$$

where $T = T_{\mu\nu}g^{\mu\nu}$, and $R_{\alpha\beta}$ is the Ricci tensor. From the right-hand side of the field equations it follows that R_{00} and R_{mn} (when $m=n$) start with order two, R_{mn} (when $m \neq n$) start with order four, while R_{0m} starts with order three. The lowest order expressions of the left-hand side are

$$R_{00} \approx \frac{1}{2} h_{00,ss},$$

$$R_{0m} \approx \frac{1}{2}\left(h_{0m,ss} - h_{0s,ms} - h_{ms,0s} + h_{ss,0m}\right), \quad (6.138)$$

$$R_{mn} \approx \frac{1}{2}\left(h_{mn,ss} - h_{ms,ns} - h_{ns,ms} - h_{00,mn} + h_{ss,mn}\right),$$

where a comma denotes a partial derivative, $\phi_{,s} = \partial_s \phi$. Hence we have

$$h_{00} = {}_2h_{00} + {}_4h_{00} + \cdots,$$

$$h_{0m} = {}_3h_{0n} + {}_5h_{0n} + \cdots, \quad (6.139)$$

$$h_{mn} = {}_2h_{mn} + {}_4h_{mn} + \cdots.$$

Newtonian Equation of Motion

We now find the equation of motion in the lowest (Newtonian) approximation. We do it in such a way as to make the generalization to the post-Newtonian approximation as simple as possible.

Because of Eqs. (6.137) and (6.138), the field equations of the lowest order are in h_{00},

$$\frac{1}{2}{}_2h_{00,ss} = \kappa\left({}_2T^{00} - \frac{1}{2}{}_2T^{00}\right) = \frac{\kappa}{2}{}_2T^{00} = \frac{\kappa}{2}\sum_{A=1}^{2}\mu_A\delta_A, \quad (6.140)$$

where, for simplicity, we have put $\mu_A = {}_2m_A$. Hence the equation obtained is

$$_2h_{00,ss} = \kappa\sum_{A=1}^{2}\mu_A\delta_A. \quad (6.141)$$

The solution of this equation that represents two masses is

$$_2h_{00} = -2G\sum_{A=1}^{2}\mu_A r_A^{-1}, \quad (6.142)$$

6.1. ELEMENTS OF GENERAL RELATIVITY

where $r_A^2 = (x^s - z_A^s)(x^s - z_A^s)$. Using $_2h_{00}$ in the equation of motion (6.132), we obtain in the lowest (second) order for the equation of motion of the first particle

$$\ddot{z}_1^k - G \int \partial_k \left(\mu_2 r_2^{-1}\right) \delta(\mathbf{x} - \mathbf{z}_1) d^3x = 0. \qquad (6.143)$$

This gives

$$\ddot{z}_1^k = G \frac{\partial}{\partial z_1^k} \frac{\mu_2}{z}, \qquad (6.144)$$

where $z^2 = (z_1^s - z_2^s)(z_1^s - z_2^s)$. Equation (6.144) is, of course, the Newtonian equation of motion.

Einstein-Infeld-Hoffmann Equation

To find the equation of motion up to the fourth order, we must know besides $_2h_{00}$ the functions $_4h_{00}$, $_3h_{0n}$ and $_2h_{mn}$. The second and third functions are easy to find. The left-hand side of the corresponding equations is written out in Eq. (6.138), whereas the right-hand side is given by Eq. (6.137) and it is $-\kappa \sum \mu_A \dot{z}_A^m \delta_A$ for the $0m$ component, and $\frac{\kappa}{2} \delta_{mn} \sum \mu_A \delta_A$ for the mn component. Therefore, for the $_2h_{mn}$ we have the equation

$$_2h_{mn,ss} - {_2h_{ms,ns}} - {_2h_{ns,ms}} + {_2h_{ss,mn}} - {_2h_{00,mn}} = \delta_{mn} {_2h_{00,ss}}, \qquad (6.145)$$

whose solution is

$$_2h_{mn} = \delta_{mn} \, _2h_{00}. \qquad (6.146)$$

The equation for $_3h_{0n}$ is

$$_3h_{0n,ss} - {_3h_{0s,ns}} - {_2h_{ns,0s}} + {_2h_{ss,n0}} = -2\kappa \sum \mu_A \dot{z}_A^n \delta_A. \qquad (6.147)$$

Using the value of $_2h_{mn}$ in terms of the $_2h_{00}$ found above, we obtain

$$_3h_{0n,ss} - {_3h_{0s,ns}} + 2 \, _2h_{00,n0} = -2\kappa \sum_{A=1}^{2} \mu_A \dot{z}_A^n \delta_A. \qquad (6.148)$$

The solution of this equation is

$$_3h_{0n} = 4G \sum_{A=1}^{2} \mu_A \dot{z}_A^n r_A^{-1}. \qquad (6.149)$$

Calculation of $_4h_{00}$ is somewhat more complicated. The relevant part of $_4h_{00}$, for two masses, that contributes to the equation of motion of the first particle, is

$$_4h_{00} \approx G\left\{2G\mu_2^2 r_2^{-2} - 3\mu_2 \dot{z}_2^s \dot{z}_2^s r_2^{-1} - \mu_2 r_{2,00} + 2G\mu_1\mu_2 (zr_2)^{-1}\right\}. \quad (6.150)$$

Using these values for $_4h_{00}$, $_3h_{0n}$, and $_2h_{mn}$ in the equation of motion (6.132) gives, for the two-body problem (Problem 6.25):

$$\ddot{z}_1^n - \mu_2 \frac{\partial(1/z)}{\partial z_1^n}$$

$$= \mu_2\bigg\{\left(\dot{z}_1^s \dot{z}_1^s + \frac{3}{2}\dot{z}_2^s \dot{z}_2^s - 4\dot{z}_1^s \dot{z}_2^s - 4\frac{\mu_2}{z} - 5\frac{\mu_1}{z}\right)\frac{\partial(1/z)}{\partial z_1^n}$$

$$+ [4\dot{z}_1^s(\dot{z}_2^n - \dot{z}_1^n) + 3\dot{z}_1^n \dot{z}_2^s - 4\dot{z}_2^n \dot{z}_2^s]\frac{\partial(1/z)}{\partial z_1^s} + \frac{1}{2}\dot{z}_2^s \dot{z}_2^r \frac{\partial^3 z}{\partial z_1^s \partial z_1^r \partial z_1^n}\bigg\}. \quad (6.151)$$

In Eq. (6.151) the Newtonian gravitational constant G was taken as equal to 1. The equation of motion for the second particle is obtained by replacing μ_1, μ_2, z_1, z_2 by μ_2, μ_1, z_2, z_1, respectively.

Equation (6.151) is known as the Einstein-Infeld-Hoffmann equation of motion, and is a generalization of the Newton equation. The essential relativistic correction may be obtained by fixing one of the particles. Writing M for μ_2, neglecting μ_1 and $\dot{z}_2^s$, and using an obvious three-dimensional vector notation, Eq. (6.151) simplifies to

$$\ddot{\mathbf{z}} - M\nabla\left(\frac{1}{z}\right) = M\left\{\left(\dot{\mathbf{z}}\cdot\dot{\mathbf{z}} - \frac{4M}{z}\right)\nabla\left(\frac{1}{z}\right) - 4\dot{\mathbf{z}}\left(\dot{\mathbf{z}}\cdot\nabla\frac{1}{z}\right)\right\}, \quad (6.152)$$

where $\mathbf{z}$ denotes the three-vector z_1^s.

6.1.8 Decomposition of the Riemann Tensor

The Riemann curvature tensor $R_{\alpha\beta\gamma\delta}$ can be decomposed into its *irreducible components*. These are the Weyl conformal tensor $C_{\alpha\beta\gamma\delta}$, the tracefree Ricci tensor $S_{\alpha\beta}$, and the Ricci scalar curvature R. The tensor $S_{\alpha\beta}$ is defined by

$$S_{\alpha\beta} = R_{\alpha\beta} - \frac{1}{4}g_{\alpha\beta}R, \quad (6.153)$$

where $R_{\alpha\beta}$ is the ordinary Ricci tensor.

6.2. THE CURVATURE SPINOR

The decomposition can be written symbolically as

$$R_{\alpha\beta\gamma\delta} = C_{\alpha\beta\gamma\delta} \bigoplus S_{\alpha\beta} \bigoplus R. \qquad (6.154)$$

No new quantities can be obtained from any of the above three irreducible components by contraction of their indices.

When written in full details, the decomposition (6.154) will then have the form:

$$R_{\rho\sigma\mu\nu} = C_{\rho\sigma\mu\nu} + \frac{1}{2}\left(g_{\rho\mu}S_{\sigma\nu} - g_{\rho\nu}S_{\sigma\mu} - g_{\sigma\mu}S_{\rho\nu} + g_{\sigma\nu}S_{\rho\mu}\right)$$

$$-\frac{1}{12}\left(g_{\rho\nu}g_{\sigma\mu} - g_{\rho\mu}g_{\sigma\nu}\right)R. \qquad (6.155)$$

It can also be written in the form:

$$R_{\rho\sigma\mu\nu} = C_{\rho\sigma\mu\nu} + \frac{1}{2}\left(g_{\rho\mu}R_{\sigma\nu} - g_{\rho\nu}R_{\sigma\mu} - g_{\sigma\mu}R_{\rho\nu} + g_{\sigma\nu}R_{\rho\mu}\right)$$

$$+\frac{1}{6}\left(g_{\rho\nu}g_{\sigma\mu} - g_{\rho\mu}g_{\sigma\nu}\right)R. \qquad (6.156)$$

6.2 The Curvature Spinor

We first derive the curvature spinor and investigate its properties. This is done along the lines of deriving the Riemann curvature tensor. Instead of applying the commutator of the covariant derivatives on a vector, however, we apply it on a spinor.

In Subsection 5.4.4 we have seen how such a procedure works when the commutator of the spin covariant derivatives were applied on a spinor. Here we use the ordinary (spacetime) covariant derivatives.

If we differentiate covariantly the quantity $\nabla_\mu \xi_Q$, given by Eq. (5.52), we obtain

$$\nabla_\nu \nabla_\mu \xi_Q = \partial_\nu \left(\nabla_\mu \xi_Q\right) - \Gamma^\lambda_{\nu\mu} \nabla_\lambda \xi_Q - \Gamma^B_{Q\nu} \nabla_\mu \xi_B, \qquad (6.157)$$

where

$$\nabla_\mu \xi_Q = \partial_\mu \xi_Q - \Gamma^P_{Q\mu} \xi_P. \qquad (6.158)$$

Substituting Eq. (6.158) in Eq. (6.157), the latter equation then yields

$$\nabla_\nu \nabla_\mu \xi_Q = \partial_\nu \partial_\mu \xi_Q - \Gamma^P_{Q\mu} \partial_\nu \xi_P - \partial_\nu \Gamma^P_{Q\mu} \xi_P - \Gamma^\lambda_{\nu\mu} \partial_\lambda \xi_Q$$

$$+\Gamma^\lambda_{\nu\mu} \Gamma^P_{Q\lambda} \xi_P - \Gamma^B_{Q\nu} \partial_\mu \xi_B + \Gamma^B_{Q\nu} \Gamma^P_{B\mu} \xi_P. \qquad (6.159)$$

Calculating now the same expression, but with the indices μ and ν being exchanged, and subtracting it from the expression (6.159), we then obtain

$$(\nabla_\nu \nabla_\mu - \nabla_\mu \nabla_\nu)\xi_Q = -F^P{}_{Q\mu\nu}\xi_P, \qquad (6.160)$$

where the mixed quantity $F^P{}_{Q\mu\nu}$ is given by

$$F^P{}_{Q\mu\nu} = \Gamma^P{}_{Q\mu,\nu} - \Gamma^P{}_{Q\nu,\mu} + \Gamma^B{}_{Q\mu}\Gamma^P{}_{B\nu} - \Gamma^B{}_{Q\nu}\Gamma^P{}_{B\mu}. \qquad (6.161)$$

In Eq. (6.161) a comma followed by a Greek letter indicates a partial differentiation, $f_{,\alpha} = \partial f/\partial x^\alpha$. The quantity $F^P{}_{Q\mu\nu}$ will be referred to in the sequel as the *curvature spinor*.

In the same way, if we apply the commutator $(\nabla_\nu \nabla_\mu - \nabla_\mu \nabla_\nu)$ to the spinor ξ^Q we then obtain

$$(\nabla_\nu \nabla_\mu - \nabla_\mu \nabla_\nu)\xi^Q = F^Q{}_{P\mu\nu}\xi^P. \qquad (6.162)$$

Equations (6.160) and (6.162) are analogous to the formulas for defining the Riemann curvature tensor (see, for instance, Eisenhart). The occurrence of the minus sign in the curvature spinor is just a matter of convention.

We may also apply the commutator of the covariant derivatives to products of spinors and spinors with more than one index, using a combination of Eqs. (6.160) and (6.162). Thus, for instance, we obtain

$$(\nabla_\nu \nabla_\mu - \nabla_\mu \nabla_\nu)(\xi_P \eta^Q) = -F^A{}_{P\mu\nu}\xi_A \eta^Q + F^Q{}_{A\mu\nu}\xi_P \eta^A, \qquad (6.163)$$

for arbitrary one-index spinors ξ_P and η^Q. Likewise we obtain

$$(\nabla_\nu \nabla_\mu - \nabla_\mu \nabla_\nu)\zeta_P{}^Q = F^Q{}_{A\mu\nu}\zeta_P{}^A - F^A{}_{P\mu\nu}\zeta_A{}^Q, \qquad (6.164)$$

for an arbitrary spinor $\zeta_P{}^Q$ with two unprimed indices.

6.2.1 Spinorial Ricci Identity

The above formulas may be further generalized to spinors of higher orders and to those with primed indices as well. Thus we obtain

$$(\nabla_\nu \nabla_\mu - \nabla_\mu \nabla_\nu)\zeta^{QS\cdots}_{PR\cdots} = F^Q{}_{A\mu\nu}\zeta^{AS\cdots}_{PR\cdots} + F^S{}_{A\mu\nu}\zeta^{QA\cdots}_{PR\cdots}$$

$$- F^A{}_{P\mu\nu}\zeta^{QS\cdots}_{AR\cdots} - F^A{}_{R\mu\nu}\zeta^{QS\cdots}_{PA\cdots}, \qquad (6.165)$$

for an arbitrary spinor $\zeta^{QS\cdots}_{PR\cdots}$ with unprimed indices. Likewise we obtain

$$(\nabla_\nu \nabla_\mu - \nabla_\mu \nabla_\nu)\zeta_{PQ'} = -F^A{}_{P\mu\nu}\zeta_{AQ'} - \overline{F}^{A'}{}_{Q'\mu\nu}\zeta_{PA'}, \qquad (6.166)$$

6.2. THE CURVATURE SPINOR

$$(\nabla_\nu \nabla_\mu - \nabla_\mu \nabla_\nu) \zeta^P_{Q'} = F^P{}_{A\mu\nu} \zeta^A_{Q'} - \overline{F}^{A'}{}_{Q'\mu\nu} \zeta^P_{A'}, \qquad (6.167)$$

$$(\nabla_\nu \nabla_\mu - \nabla_\mu \nabla_\nu) \zeta^{PQ'} = F^P{}_{A\mu\nu} \zeta^{AQ'} + \overline{F}^{Q'}{}_{A'\mu\nu} \zeta^{PA'}, \qquad (6.168)$$

for the arbitrary spinors $\zeta_{PQ'}$, $\zeta^P_{Q'}$ and $\zeta^{PQ'}$ with mixed indices. Equation (6.165) will be referred to as the *Spinorial Ricci identity*.

6.2.2 Symmetry of the Curvature Spinor

We now study the symmetry properties of the curvature spinor introduced above. Later on we will relate it to the Riemann curvature tensor.

From Eq. (6.160) we obtain

$$(\nabla_\nu \nabla_\mu - \nabla_\mu \nabla_\nu) \xi_Q = F_{PQ\mu\nu} \xi^P. \qquad (6.169)$$

We may, on the other hand, lower the free index Q in Eq. (6.162), thus getting

$$(\nabla_\nu \nabla_\mu - \nabla_\mu \nabla_\nu) \xi_Q = F_{QP\mu\nu} \xi^P. \qquad (6.170)$$

Comparing now the last two equations we find that the curvature spinor satisfies the property

$$F_{PQ\mu\nu} = F_{QP\mu\nu}, \qquad (6.171)$$

namely, it is symmetric with respect to its two spinor indices P and Q. By its definition, furthermore, it is antisymmetric in its spacetime tensorial indices μ and ν, namely

$$F_{PQ\mu\nu} = -F_{PQ\nu\mu}. \qquad (6.172)$$

To further study the symmetry properties of the curvature spinor we define the spinor

$$F_{PQAB'CD'} = F_{PQ\mu\nu} \sigma^\mu_{AB'} \sigma^\nu_{CD'}, \qquad (6.173)$$

which is, of course, skew-symmetric under the exchange of the pairs of indices AB' and CD'. Hence it may be decomposed, similar to the spinor equivalent of the electromagnetic field tensor given by Eq. (5.83).

Accordingly we may write

$$F_{PQAB'CD'} = -(\chi_{PQAC}\epsilon_{B'D'} + \phi_{PQB'D'}\epsilon_{AC}), \qquad (6.174)$$

where the minus sign is introduced for convenience, and where the two new spinors χ_{PQAC} and $\phi_{PQB'D'}$ are defined by

$$\chi_{PQAC} = -\frac{1}{2} F_{PQAB'C}{}^{B'}, \qquad (6.175)$$

$$\phi_{PQB'D'} = -\frac{1}{2} F_{PQAB'}{}^{A}{}_{D'}. \qquad (6.176)$$

In the following we study the properties of the above two spinors. Before doing so we relate the curvature spinor to the Riemann curvature tensor. As we see, the curvature spinor has only six spinorial indices. Since the Riemann tensor has four spacetime indices, its spinor equivalent will have eight spinorial indices. We will see in the next section that the two spinors χ and ϕ describe completely the Riemann spinor. It thus follows that the curvature of spacetime is determined by a six-indices spinor (the curvature spinor) and not by an eight-indices spinor (the Riemann spinor). In fact, we will see that the Riemann spinor is obtained from the curvature spinor and its complex conjugate.

6.3 Relation to the Riemann Tensor

Multiplying Eq. (6.166) by $\sigma_\alpha^{PQ'}$ and rearranging the indices we obtain

$$(\nabla_\nu \nabla_\mu - \nabla_\mu \nabla_\nu)\zeta_{PQ'}\sigma_\alpha^{PQ'} = -\left(F^C{}_{P\mu\nu}\sigma_\alpha^{PD'} + \overline{F}^{D'}{}_{Q'\mu\nu}\sigma_\alpha^{CQ'}\right)\zeta_{CD'}. \qquad (6.177)$$

Hence we may write, since $\zeta_{PQ'}\sigma_\alpha^{PQ'} = \zeta_\alpha$ is a vector,

$$(\nabla_\nu \nabla_\mu - \nabla_\mu \nabla_\nu)\zeta_\alpha = R^{CD'}{}_{\alpha\mu\nu}\zeta_{CD'} = R^\rho{}_{\alpha\mu\nu}\zeta_\rho, \qquad (6.178)$$

where use has been made of the notation

$$R^{CD'}{}_{\alpha\mu\nu} = -\left(F^C{}_{P\mu\nu}\sigma_\alpha^{PD'} + \overline{F}^{D'}{}_{Q'\mu\nu}\sigma_\alpha^{CQ'}\right), \qquad (6.179)$$

and

$$R^\rho{}_{\alpha\mu\nu} = R^{CD'}{}_{\alpha\mu\nu}\sigma^\rho_{CD'}, \qquad (6.180)$$

by Eq. (6.178). The tensor given by Eq. (6.180) is the Riemann curvature tensor. The last two formulas give the relationship between the curvature spinor and its complex conjugate on the one hand, and the Riemann curvature tensor on the other hand.

From Eq. (6.179) we now obtain

$$R_{AB'EF'\mu\nu} = R^{CD'}{}_{\alpha\mu\nu}\epsilon_{CA}\epsilon_{D'B'}\sigma^\alpha_{EF'}$$

$$= F_{AE\mu\nu}\epsilon_{B'F'} + \epsilon_{AE}\overline{F}_{B'F'\mu\nu}. \qquad (6.181)$$

6.3. RELATION TO THE RIEMANN TENSOR

Equivalently, the latter equation may be written in the form

$$R_{AB'EF'MN'PQ'} = F_{AEMN'PQ'}\epsilon_{B'F'} + \epsilon_{AE}\overline{F}_{B'F'N'MQ'P}. \tag{6.182}$$

The left-hand side of the above formula is the spinor equivalent of the Riemann curvature tensor.

We may also obtain equations for the curvature tensor. Multiplying Eq. (6.181) by $\epsilon^{B'F'}$ we obtain

$$F_{AE\mu\nu} = \frac{1}{2}R_{AB'E}{}^{B'}{}_{\mu\nu} = \frac{1}{2}\sigma_{\alpha AB'}\sigma^{\beta}{}_{E}{}^{B'}R^{\alpha}{}_{\beta\mu\nu}. \tag{6.183}$$

Likewise we obtain

$$\overline{F}_{B'F'\mu\nu} = \frac{1}{2}R_{AB'}{}^{A}{}_{F'\mu\nu} = \frac{1}{2}\sigma_{\alpha AB'}\sigma^{\beta A}{}_{F'}R^{\alpha}{}_{\beta\mu\nu}, \tag{6.184}$$

by multiplying Eq. (6.181) by ϵ^{AE}.

6.3.1 Bianchi Identities

We may also write the Bianchi identities in terms of the curvature spinor. From Eq. (6.183) we then obtain

$$\nabla_\alpha F_{AE\beta\gamma} + \nabla_\beta F_{AE\gamma\alpha} + \nabla_\gamma F_{AE\alpha\beta}$$
$$= \frac{1}{2}\sigma_{\mu AB'}\sigma^{\nu}{}_{E}{}^{B'}\left(\nabla_\alpha R^{\mu}{}_{\nu\beta\gamma} + \nabla_\beta R^{\mu}{}_{\nu\gamma\alpha} + \nabla_\gamma R^{\mu}{}_{\nu\alpha\beta}\right) = 0, \tag{6.185}$$

where the last equality is obtained from the ordinary (tensorial) Bianchi identities,

$$\nabla_\gamma R^{\mu}{}_{\nu\alpha\beta} + \nabla_\beta R^{\mu}{}_{\nu\gamma\alpha} + \nabla_\alpha R^{\mu}{}_{\nu\beta\gamma} = 0. \tag{6.186}$$

Defining the dual to the curvature spinor by

$${}^\star F_{PQ}{}^{\alpha\beta} = \frac{1}{2\sqrt{-g}}\epsilon^{\alpha\beta\mu\nu}F_{PQ\mu\nu} = \frac{1}{2}F_{PQ}{}^{\mu\nu}\epsilon_{\mu\nu}{}^{\alpha\beta}, \tag{6.187}$$

the Bianchi identities may then be written in the form

$$\nabla_\beta {}^\star F_{PQ}{}^{\alpha\beta} = 0. \tag{6.188}$$

The relations (6.185) and (6.188) may also be written in the forms

$$\nabla_{AB'}F_{PQCD'EF'} + \nabla_{CD'}F_{PQEF'AB'} + \nabla_{EF'}F_{PQAB'CD'} = 0, \tag{6.189}$$

and

$$\nabla^{GH'}{}^\star F_{PQEF'GH'} = 0, \tag{6.190}$$

respectively, when written in spinor forms.

In the next section we further discuss the gravitational field dynamical variables.

6.4 The Gravitational Field Spinors

We are now in a position to find the spinors in terms of which the gravitational field is described. We have already found in Chapter 5 the spinor equivalent to the geometrical metric tensor $g_{\mu\nu}$ whose expression was shown to be given by the flat spacetime metric

$$g_{AB'CD'} = \epsilon_{AC}\epsilon_{B'D'} = \begin{pmatrix} 0 & & & 0 & 1 \\ & & & -1 & 0 \\ 0 & -1 & & & \\ 1 & 0 & & & 0 \end{pmatrix}. \quad (6.191)$$

The rows and the columns of the 4×4 matrix (6.191) are labeled by the pairs of indices AB' and CD', each taking the values $(1, 2, 3, 4)=(00', 01', 10', 11')$.

6.4.1 Decomposition of the Riemann Tensor

We next discuss the Riemann curvature tensor and decompose its spinor equivalent, which is given by

$$R_{AB'CD'EF'GH'} = \sigma^\alpha_{AB'}\sigma^\beta_{CD'}\sigma^\gamma_{EF'}\sigma^\delta_{GH'}R_{\alpha\beta\gamma\delta}. \quad (6.192)$$

Decomposing this spinor by the method of decomposing the spinor equivalent of the electromagnetic field used in Chapter 5, we then obtain

$$R_{AB'CD'EF'GH'} = \frac{1}{2}\left(\epsilon_{AC}R_{PB'}{}^P{}_{D'EF'GH'} + R_{AP'C}{}^{P'}{}_{EF'GH'}\epsilon_{B'D'}\right)$$

$$= \frac{1}{4}\epsilon_{AC}\left(R_{PB'}{}^P{}_{D'KF'}{}^K{}_{H'}\epsilon_{EG} + R_{PB'}{}^P{}_{D'EL'G}{}^{L'}\epsilon_{F'H'}\right)$$

$$+ \frac{1}{4}\left(R_{AP'C}{}^{P'}{}_{KF'}{}^K{}_{H'}\epsilon_{EG} + R_{AP'C}{}^{P'}{}_{EL'G}{}^{L'}\epsilon_{F'H'}\right)\epsilon_{B'D'}. \quad (6.193)$$

The proof of the above formula is left for the reader (see Problem 6.3).

To compare the above decomposition for the spinor equivalent of the curvature tensor with that given in the last section for the same tensor, we denote the last two terms on the right-hand side of Eq. (6.193) as follows:

$$\chi_{ACEG} = -\frac{1}{4}R_{AP'C}{}^{P'}{}_{EL'G}{}^{L'}, \quad (6.194)$$

$$\phi_{ACF'H'} = -\frac{1}{4}R_{AP'C}{}^{P'}{}_{KF'}{}^K{}_{H'}. \quad (6.195)$$

6.4. THE GRAVITATIONAL FIELD SPINORS

Using now the decomposition (6.182) for the spinor equivalent to the Riemann curvature tensor in Eqs. (6.194) and (6.195), we get

$$\chi_{ACEG} = -\frac{1}{2} F_{ACEL'G}{}^{L'}, \tag{6.196}$$

$$\phi_{ACF'H'} = -\frac{1}{2} F_{ACKF'}{}^{K}{}_{H'}. \tag{6.197}$$

Comparing the last two formulas with Eqs. (6.175) and (6.176), we find that they are identical. Hence Eqs. (6.194) and (6.195) are consistent with our previous definitions for the same quantities χ_{PQAB} and $\phi_{PQA'B'}$ given by Eqs. (6.175) and (6.176), respectively, when the decomposition (6.182) is used.

The decomposition of the spinor equivalent of the Riemann curvature tensor, given by Eq. (6.193), may be further simplified if we notice that the first two terms on the right-hand side of that equation may be written in terms of the complex conjugate of the spinors χ_{ABCD} and $\phi_{ABC'D'}$. To see this we use the fact that the Riemann curvature tensor is real, and therefore it satisfies

$$R_{PB'}{}^{P}{}_{D'KF'}{}^{K}{}_{H'} = \overline{R}_{B'PD'}{}^{P}{}_{F'KH'}{}^{K} = \overline{R_{BP'D}{}^{P'}{}_{FK'H}{}^{K'}}$$

$$= -4\overline{\chi_{BDFH}} = -4\overline{\chi}_{B'D'F'H'}, \tag{6.198}$$

$$R_{PB'}{}^{P}{}_{D'EL'G}{}^{L'} = \overline{R}_{B'PD'}{}^{P}{}_{L'E}{}^{L'}{}_{G} = \overline{R_{BP'D}{}^{P'}{}_{LE'}{}^{L}{}_{G'}}$$

$$= -4\overline{\phi_{BDE'G'}} = -4\overline{\phi}_{B'D'EG}, \tag{6.199}$$

by Eqs. (6.194) and (6.195). Accordingly we finally obtain

$$R_{AB'CD'EF'GH'} = -(\chi_{ACEG}\,\epsilon_{B'D'}\,\epsilon_{F'H'} + \phi_{ACF'H'}\,\epsilon_{B'D'}\,\epsilon_{EG}$$

$$+ \epsilon_{AC}\,\overline{\phi}_{B'D'EG}\,\epsilon_{F'G'} + \epsilon_{AC}\,\epsilon_{EG}\,\overline{\chi}_{B'D'F'H'}), \tag{6.200}$$

for the decomposition of the spinor equivalent to the Riemann curvature tensor.

We next decompose the spinor equivalent to the dual of the curvature tensor. If $^*R_{\alpha\beta\gamma\delta}$ is the dual to the Riemann curvature tensor, then its spinor equivalent is given by

$$^*R_{AB'CD'EF'GH'} = i(\chi_{ACEG}\,\epsilon_{B'D'}\,\epsilon_{F'H'} - \phi_{ACF'H'}\,\epsilon_{B'D'}\,\epsilon_{EG}$$

$$+ \epsilon_{AC}\,\overline{\phi}_{B'D'EG}\,\epsilon_{F'G'} - \epsilon_{AC}\,\epsilon_{EG}\,\overline{\chi}_{B'D'F'H'}), \tag{6.201}$$

The proof of the above formula is given in Problem 6.4.

6.4.2 The Gravitational Spinor

The two spinors χ_{ABCD} and $\phi_{ABC'D'}$ uniquely determine the spinor equivalent to the Riemann curvature tensor. The symmetry properties of the spinor χ_{ABCD} follow from the symmetry properties of the Riemann tensor given by

$$R_{\alpha\beta\gamma\delta} = -R_{\beta\alpha\gamma\delta} = -R_{\alpha\beta\delta\gamma}, \tag{6.202a}$$

$$R_{\alpha\beta\gamma\delta} = R_{\gamma\delta\alpha\beta}, \tag{6.202b}$$

$$R_{\alpha\beta\gamma\delta} + R_{\alpha\gamma\delta\beta} + R_{\alpha\delta\beta\gamma} = 0. \tag{6.202c}$$

Because of the relation $R_{\alpha\beta\gamma\delta} = -R_{\beta\alpha\gamma\delta}$, for instance, we have

$$\chi_{CAEG} = -\frac{1}{4} R_{CP'A}{}^{P'}{}_{EL'G}{}^{L'} = \frac{1}{4} R_A{}^{P'}{}_{CP'EL'G}{}^{L'}$$

$$= -\frac{1}{4} R_{AP'C}{}^{P'}{}_{EL'G}{}^{L'} = \chi_{ACEG}. \tag{6.203}$$

In the same way, using the fact that $R_{\alpha\beta\gamma\delta} = -R_{\alpha\beta\delta\gamma}$, we find that χ_{ACEG} is symmetric with respect to the two indices E and G, namely, $\chi_{ACEG} = \chi_{ACGE}$. Finally, using the fact that $R_{\alpha\beta\gamma\delta} = R_{\gamma\delta\alpha\beta}$ leads to the symmetry of χ_{ACEG} under the exchange of the first and second pairs of indices, $\chi_{ACEG} = \chi_{EGAC}$. Accordingly we have

$$\chi_{ABCD} = \chi_{BACD} = \chi_{ABDC}, \tag{6.204}$$

$$\chi_{ABCD} = \chi_{CDAB}, \tag{6.205}$$

which the spinor χ_{ABCD} satisfies.

Similarly we find that the spinor $\phi_{ABC'D'}$ is symmetric under the exchange of indices A and B and C' and D'. We have, moreover,

$$\overline{\phi}_{C'D'AB} = \overline{\phi_{CDA'B'}} = -\frac{1}{4}\overline{R_{CP'D}{}^{P'}{}_{KA'}{}^{K}{}_{B'}}$$

$$= -\frac{1}{4}\overline{R_{C'PD'}{}^{P}{}_{K'A}{}^{K'}{}_{B}} = -\frac{1}{4}\overline{R_{K'A}{}^{K'}{}_{BC'PD'}{}^{P}}, \tag{6.206}$$

by Eq. (6.195) and using the symmetry of the Riemann tensor. Using now the fact that the Riemann tensor is real and hence its spinor equivalent is Hermitian, we then obtain

$$\overline{\phi}_{C'D'AB} = -\frac{1}{4} R_{AK'B}{}^{K'}{}_{PC'}{}^{P}{}_{D'} = \phi_{ABC'D'}. \tag{6.207}$$

6.4. THE GRAVITATIONAL FIELD SPINORS

Summarizing the above results we find the following formulas:

$$\phi_{ABC'D'} = \phi_{BAC'D'} = \phi_{ABD'C'}, \qquad (6.208)$$

$$\phi_{ABC'D'} = \overline{\phi}_{C'D'AB}. \qquad (6.209)$$

Equations (6.208) and (6.209) express the symmetry properties of the spinor $\phi_{ABC'D'}$.

Because of the symmetry properties (6.204) and (6.205), the spinor χ_{ABCD} behaves like a 3×3 symmetric complex matrix. This fact may easily be seen since each pair of the indices AB and CD takes the three values $(1, 2, 3)=(00, 01=10, 11)$ and χ_{ABCD} is unchanged under the exchange of AB with CD. Hence the spinor χ_{ABCD} may have at most six complex components. These components, however, are not entirely independent. For if we calculate the trace of χ_{ABCD},

$$\lambda = \chi_{AB}{}^{AB} = \epsilon^{AC} \epsilon^{BD} \chi_{ABCD}, \qquad (6.210)$$

we find that λ is a real quantity.

Reality of λ

The reality of λ may be seen using the symmetry property expressed by $*R^\rho{}_{\alpha\rho\beta} = 0$, which the dual to the Riemann curvature tensor satisfies. In spinor calculus the above equation is given by

$$*R^{EF'}{}_{AB'EF'CD'} = 0. \qquad (6.211)$$

Using the expression for the spinor equivalent to the dual of the Riemann curvature tensor, given by Eq. (6.201), in Eq. (6.211), we then find

$$*R^{EF'}{}_{AB'EF'CD'}$$
$$= i\left(-\chi^E{}_{AEC}\,\epsilon_{B'D'} + \phi_{CAB'D'} - \overline{\phi}_{D'B'AC} + \epsilon_{AC}\,\overline{\chi}^{F'}{}_{B'F'D'}\right) = 0. \qquad (6.212)$$

The two terms with ϕ and $\overline{\phi}$ on the right-hand side of the above formula cancel out because of Eqs. (6.208) and (6.209). Hence Eq. (6.212) reduces to

$$\chi^E{}_{AEC}\,\epsilon_{B'D'} = \epsilon_{AC}\,\overline{\chi}^{F'}{}_{B'F'D'}, \qquad (6.213)$$

from which, by multiplying it by $\epsilon^{D'B'}$,

$$\chi^E{}_{AEC} = -\frac{1}{2}\epsilon_{AC}\,\overline{\chi}^{F'D'}{}_{F'D'} = -\frac{1}{2}\epsilon_{AC}\,\overline{\chi}_{F'D'}{}^{F'D'}. \qquad (6.214)$$

Accordingly we obtain

$$\chi_{EA}{}^E{}_C = \frac{1}{2}\epsilon_{AC}\overline{\lambda}, \qquad (6.215)$$

where λ is defined by Eq. (6.210). Multiplying now Eq. (6.215) by ϵ^{AC}, the latter equation then yields

$$\lambda = \overline{\lambda}, \qquad (6.216)$$

namely, λ is real.

As a consequence of the reality of λ, the spinor χ_{ABCD} has only 11 independent real components rather then 12. In the sequel the spinor χ_{ABCD} is shown to describe the Weyl spinor plus the Ricci scalar curvature, and it will be referred to as the *gravitational spinor*.

The spinor $\phi_{ABC'D'}$, on the other hand, behaves like a 3 × 3 Hermitian matrix. This fact may easily be seen if we write $\phi_{ABC'D'}$ in the form of the matrix

$$\Phi = \begin{pmatrix} \phi_{00} & \phi_{01} & \phi_{02} \\ \phi_{10} & \phi_{11} & \phi_{12} \\ \phi_{20} & \phi_{21} & \phi_{22} \end{pmatrix} = \begin{pmatrix} \phi_{000'0'} & \phi_{000'1'} & \phi_{001'1'} \\ \phi_{010'0'} & \phi_{010'1'} & \phi_{011'1'} \\ \phi_{110'0'} & \phi_{110'1'} & \phi_{111'1'} \end{pmatrix}. \qquad (6.217)$$

Hence the matrix elements satisfy $\phi_{mn} = \overline{\phi}_{nm}$, with $m, n = 0, 1, 2$, by Eq. (6.209), namely, the matrix Φ is Hermitian, $\Phi^\dagger = \Phi$.

Accordingly the spinor $\phi_{ABC'D'}$ has three complex components ϕ_{01}, ϕ_{02}, ϕ_{12} and three real components ϕ_{00}, ϕ_{11}, ϕ_{22}, namely, it has nine real independent components. In the sequel the spinor $\phi_{ABC'D'}$ is shown to describe the tracefree Ricci tensor

$$S_{\mu\nu} = R_{\mu\nu} - \frac{1}{4}g_{\mu\nu}R. \qquad (6.218)$$

6.4.3 The Ricci Spinor

We now calculate the Ricci spinor. It is given by

$$R_{AB'CD'} = R^{EF'}{}_{AB'EF'CD'} = \epsilon^{EG}\epsilon^{F'H'}R_{GH'AB'EF'CD'}. \qquad (6.219)$$

Using the expression (6.200) for the spinor equivalent to the Riemann curvature tensor, we then obtain

$$R_{AB'CD'} = -\left(\chi_{EA}{}^E{}_C\,\epsilon_{B'D'} - \phi_{CAB'D'} - \overline{\phi}_{D'B'AC} + \epsilon_{AC}\,\overline{\chi}_{F'B'}{}^{F'}{}_{D'}\right). \qquad (6.220)$$

6.4. THE GRAVITATIONAL FIELD SPINORS

The second and third terms on the right-hand side of the above formula are equal to each other by Eqs. (6.208) and (6.209). Moreover, from Eqs. (6.215) and (6.216) we obtain

$$\chi_{EA}{}^{E}{}_{C} = \frac{1}{2}\epsilon_{AC}\lambda, \tag{6.221a}$$

$$\overline{\chi}_{F'B'}{}^{F'}{}_{D'} = \frac{1}{2}\epsilon_{B'D'}\lambda, \tag{6.221b}$$

Hence the Ricci spinor is given by

$$R_{AB'CD'} = 2\phi_{ACB'D'} - \lambda\,\epsilon_{AC}\,\epsilon_{B'D'}. \tag{6.222}$$

The Ricci scalar curvature is given by $R = R^{\mu}{}_{\mu}$. Hence we have

$$R = R_{AB'}{}^{AB'} = -4\lambda. \tag{6.223}$$

As a result, the spinor equivalent to the tracefree Ricci tensor is given by

$$S_{AB'CD'} = R_{AB'CD'} - \frac{1}{4}g_{AB'CD'}R, \tag{6.224}$$

or, using Eqs. (6.222) and (6.223), we obtain

$$S_{AB'CD'} = 2\phi_{ACB'D'}. \tag{6.225}$$

Hence the spinor $\phi_{ACB'D'}$ is equal to one half the spinor equivalent of the tracefree Ricci tensor.

The spinor equivalent of the Einstein tensor is given by

$$G_{AB'CD'} = R_{AB'CD'} - \frac{1}{2}g_{AB'CD'}R$$

$$= 2\phi_{ACB'D'} + \lambda\,\epsilon_{AC}\,\epsilon_{B'D'}. \tag{6.226}$$

6.4.4 The Weyl Spinor

The decomposition of the spinor equivalent to the Riemann tensor given above is not complete since the spinor χ_{ABCD} may be further decomposed and related to the Weyl conformal spinor. To this end we write the spinor χ_{ABCD} in the form

$$\chi_{ABCD} = \frac{1}{3}\left(\chi_{ABCD} + \chi_{ACBD} + \chi_{ADBC}\right) + \frac{1}{3}\left(\chi_{ABCD} - \chi_{ACBD}\right)$$

$$+\frac{1}{3}\left(\chi_{ABCD}-\chi_{ADBC}\right). \tag{6.227}$$

Hence we may write

$$\chi_{ABCD}=\psi_{ABCD}+\frac{1}{3}\left(\chi_{ABCD}-\chi_{ACBD}\right)+\frac{1}{3}\left(\chi_{ABCD}-\chi_{ADBC}\right), \tag{6.228}$$

where

$$\psi_{ABCD}=\frac{1}{3}\left(\chi_{ABCD}+\chi_{ACBD}+\chi_{ADBC}\right). \tag{6.229}$$

We notice that the first expression in brackets on the right-hand side of Eq. (6.228) is antisymmetric in the indices B and C. Hence using Eq. (5.70), it may be written in the form

$$\frac{1}{3}\left(\chi_{ABCD}-\chi_{ACBD}\right)=\frac{1}{3}\chi_{AE}{}^{E}{}_{D}\,\epsilon_{BC}. \tag{6.230}$$

The last term of Eq. (6.228) may also be written as

$$\frac{1}{3}\left(\chi_{ABCD}-\chi_{ADBC}\right)=\frac{1}{3}\left(\chi_{ABDC}-\chi_{ADBC}\right)=\frac{1}{3}\chi_{AE}{}^{E}{}_{C}\,\epsilon_{BD}, \tag{6.231}$$

by Eqs. (6.204) and (5.70). Using now Eq. (6.221), furthermore, we finally obtain the following for Eq. (6.228):

$$\chi_{ABCD}=\psi_{ABCD}+\frac{\lambda}{6}\left(\epsilon_{AC}\epsilon_{BD}+\epsilon_{AD}\epsilon_{BC}\right). \tag{6.232}$$

Since λ is a real quantity, it follows that the spinor ψ_{ABCD} has 10 independent real components.

The symmetry of the spinor ψ_{ABCD} may be found as follows. From Eq. (6.232) we see that it satisfies the same symmetry as the spinor χ_{ABCD}, namely, $\psi_{ABCD}=\psi_{BACD}=\psi_{ABDC}=\psi_{CDAB}$. In fact the spinor ψ_{ABCD} is symmetric with respect to all of its four indices. For instance

$$\psi_{ACBD}=\frac{1}{3}\left(\chi_{ACBD}+\chi_{ABCD}+\chi_{ADCB}\right), \tag{6.233}$$

by definition. Hence

$$\psi_{ACBD}=\frac{1}{3}\left(\chi_{ABCD}+\chi_{ACBD}+\chi_{ADBC}\right)=\psi_{ABCD}, \tag{6.234}$$

where use has been made of $\chi_{ADCB}=\chi_{ADBC}$.

The totally symmetric spinor ψ_{ABCD} has thus only five independent complex components, as has been pointed out above. These components

6.4. THE GRAVITATIONAL FIELD SPINORS

are ψ_{0000}, ψ_{0001}, ψ_{0011}, ψ_{0111}, and ψ_{1111}. These components are sometimes denoted as follows:

$$\psi_0 = \psi_{0000}, \quad \psi_1 = \psi_{0001}, \quad \psi_2 = \psi_{0011}, \quad \psi_3 = \psi_{0111}, \quad \psi_4 = \psi_{1111}. \tag{6.235}$$

Using now the decomposition (6.228) for the gravitational spinor χ_{ABCD} we may then find the decomposition of the spinor equivalent to the Riemann tensor given by Eq. (6.200). We obtain

$$R_{AB'CD'EF'GH'} = -(\psi_{ACEG}\, \epsilon_{B'D'}\, \epsilon_{F'H'} + \phi_{ACF'H'}\, \epsilon_{B'D'}\, \epsilon_{EG}$$
$$+ \epsilon_{AC}\, \overline{\phi}_{B'D'EG}\, \epsilon_{F'H'} + \epsilon_{AC}\, \epsilon_{EG}\, \overline{\psi}_{B'D'F'H'})$$
$$- \frac{\lambda}{6}[(\epsilon_{AE}\epsilon_{CG} + \epsilon_{AG}\epsilon_{CE})\, \epsilon_{B'D'}\, \epsilon_{F'H'}$$
$$+ \epsilon_{AC}\, \epsilon_{EG}\, (\epsilon_{B'F'}\, \epsilon_{D'H'} + \epsilon_{B'H'}\, \epsilon_{D'F'})]. \tag{6.236}$$

Relation to the Weyl Tensor

We now show the relationship between the spinor ψ_{ABCD} and the spinor equivalent to the Weyl conformal tensor $C_{\alpha\beta\gamma\delta}$. Let us denote the latter spinor by $C_{AB'CD'EF'GH'}$. We show below that

$$C_{AB'CD'EF'GH'} = -(\psi_{ACEG}\, \epsilon_{B'D'}\, \epsilon_{F'H'} + \epsilon_{AC}\, \epsilon_{EG}\overline{\psi}_{B'D'F'H'}). \tag{6.237}$$

From Eq. (6.236) we see that the spinor (6.237) satisfies the same symmetries as those of the spinor equivalent to the Riemann curvature tensor. We have to show, in addition, that the trace

$$C^{EF'}{}_{CD'EF'GH'} = 0. \tag{6.238}$$

Indeed a direct calculation verifies that Eq. (6.238) is satisfied.

Hence the spinor ψ_{ABCD} is equivalent to the Weyl conformal tensor and is referred to in the sequel as the *Weyl conformal spinor*.

If $C_{\alpha\beta\gamma\delta}$ is the Weyl conformal tensor and $^*C_{\alpha\beta\gamma\delta}$ is its dual,

$$^*C_{\alpha\beta\gamma\delta} = \frac{1}{2}\sqrt{-g}C_{\alpha\beta}{}^{\mu\nu}\epsilon_{\mu\nu\gamma\delta}, \tag{6.239}$$

then the spinor equivalent to the tensor

$$C^+_{\alpha\beta\gamma\delta} = C_{\alpha\beta\gamma\delta} + i{}^*C_{\alpha\beta\gamma\delta} \tag{6.240}$$

is given by
$$C^+_{AB'CD'EF'GH'} = -2\psi_{ACEG}\,\epsilon_{B'D'}\,\epsilon_{F'H'}. \tag{6.241}$$
The proof of the above formula is given in Problem 6.5.

As a consequence of the decomposition of the gravitational spinor ψ_{ABCD} into the Weyl conformal spinor plus the Ricci scalar curvature, the curvature spinor (6.174) may finally be written in the form
$$F_{PQAB'CD'} = -\left[\psi_{PQAC} + \frac{\lambda}{6}(\epsilon_{PA}\,\epsilon_{QC} + \epsilon_{PC}\,\epsilon_{QA})\right]\epsilon_{B'D'} - \phi_{PQB'D'}\,\epsilon_{AC}. \tag{6.242}$$
Equation (6.242) describes the decomposition of the curvature spinor into its irreducible components, namely, the Weyl spinor, the tracefree Ricci spinor, and the Ricci scalar curvature. This is similar to the decomposition of the Riemann curvature tensor into its irreducible components (see Subsection 6.1.8).

In analogy with the curvature spinor we may define the *conformal spinor* by
$$\psi_{PQ\alpha\beta} = \psi_{PQAB}\sigma_\alpha{}^A{}_{C'}\sigma_\beta{}^{BC'}. \tag{6.243}$$
Under the conformal transformation $\tilde{g}_{\mu\nu} = e^{2\beta}g_{\mu\nu}$, the matrices σ_μ transform into $\tilde{\sigma}_\mu$ given by
$$\tilde{\sigma}_\mu(x) = e^\beta \sigma_\mu(x), \tag{6.244a}$$
$$\tilde{\sigma}^\mu(x) = e^{-\beta}\sigma^\mu(x), \tag{6.244b}$$
We now find the transformed components of the conformal spinor $\tilde{\psi}_{PQ\alpha\beta}$ under the conformal transformation. From Eq. (6.241) we find that
$$\psi_{PQAB} = -\frac{1}{8}C^+_{PC'Q}{}^{C'}{}_{AF'B}{}^{F'}. \tag{6.245}$$
Hence the conformal spinor, by Eq. (6.243), is given by
$$\psi_{PQ\alpha\beta} = -\frac{1}{8}C^+_{PC'Q}{}^{C'}{}_{AF'B}{}^{F'}\sigma_\alpha{}^A{}_{D'}\sigma_\beta{}^{BD'}$$
$$= -\frac{1}{8}C^{+\mu}{}_{\nu\kappa\lambda}\sigma_{\mu PC'}\sigma^\nu{}_Q{}^{C'}\sigma^\kappa{}_{AF'}\sigma^\lambda{}_B{}^{F'}\sigma_\alpha{}^A{}_{D'}\sigma_\beta{}^{BD'}. \tag{6.246}$$
Since $\tilde{C}^{+\mu}{}_{\nu\kappa\lambda} = C^{+\mu}{}_{\nu\kappa\lambda}$, and because the expression (6.246) includes an equal number of terms of the matrices σ^α having covariant and contravariant spacetime tensor indices, we find that
$$\tilde{\psi}_{PQ\alpha\beta} = \psi_{PQ\alpha\beta}. \tag{6.247}$$
Therefore the Weyl spinor is invariant under the conformal transformation.

6.4.5 The Bianchi Identities

The Bianchi identities may be written in terms of the dual to the Riemann tensor in the form given by

$$\nabla^{\rho\star} R_{\alpha\beta\gamma\rho} = 0. \tag{6.248}$$

The spinor equivalent to this equation is given by

$$\nabla^{GH'\star} R_{AB'CD'EF'GH'} = 0. \tag{6.249}$$

Here the covariant differentiation operator $\nabla^{GH'}$ is defined by Eq. (5.61).

Using the expression (6.201), for the spinor equivalent of the dual to the Riemann curvature tensor, in Eq. (6.249) we obtain

$$\nabla^G_{F'} \chi_{ACEG} \, \epsilon_{B'D'} - \nabla^{H'}_E \phi_{ACF'H'} \, \epsilon_{B'D'} + \epsilon_{AC} \nabla^G_{F'} \overline{\phi}_{B'D'EG}$$

$$-\epsilon_{AC} \nabla^{H'}_E \overline{\chi}_{B'D'F'H'} = 0. \tag{6.250}$$

Multiplying the above equation by $\epsilon^{B'D'}$ then gives

$$\nabla^G_{F'} \chi_{ACEG} - \nabla^{H'}_E \phi_{ACF'H'} = 0. \tag{6.251}$$

Equations (6.251) are the Bianchi identities in spinor calculus.

In the next chapter the spinors of gauge fields are discussed.

6.5 Problems

6.1 Find the expressions for the differential operators

$$\nabla_{(AC)} = \frac{1}{2} \left(\nabla_{AE'} \nabla_C^{E'} + \nabla_{CE'} \nabla_A^{E'} \right), \tag{1}$$

$$\nabla_{(B'D')} = \frac{1}{2} \left(\nabla_{EB'} \nabla^E_{D'} + \nabla_{ED'} \nabla^E_{B'} \right), \tag{2}$$

when applied on an arbitrary unprimed one-index spinor ξ_Q.

Solution: By Eqs. (6.170) and (6.171) we have

$$(\nabla_{CD'} \nabla_{AB'} - \nabla_{AB'} \nabla_{CD'}) \xi_Q = F_{PQAB'CD'} \xi^P. \tag{3}$$

Using the decomposition of the commutator operator given in Problem 5.1, and using the decomposition of the curvature spinor given by Eq. (6.174), we then obtain

$$\left[\epsilon_{D'B'} \nabla_{(AC)} + \epsilon_{CA} \nabla_{(B'D')} \right] \xi_Q = - \left(\chi_{PQAC} \, \epsilon_{B'D'} + \phi_{PQB'D'} \epsilon_{AC} \right) \xi^P. \tag{4}$$

Multiplying now the latter equation by $\epsilon^{D'B'}$ and ϵ^{CA} we obtain

$$\nabla_{(AC)}\xi_Q = \chi_{PQAC}\xi^P, \qquad \nabla_{(B'D')}\xi_Q = \phi_{PQB'D'}\xi^P, \qquad (5)$$

In the same way we obtain the corresponding results when the operators (1) and (2) apply on a primed-index spinor $\eta_{Q'}$. We obtain

$$\nabla_{(AC)}\eta_{Q'} = \overline{\phi}_{P'Q'AC}\eta^{P'}, \qquad \nabla_{(B'D')}\eta_{Q'} = \overline{\chi}_{P'Q'B'D'}\eta^{P'}. \qquad (6)$$

6.2 Show that

$$\nabla_{(AB)}\zeta_Q^{DE'} = \chi_{PQAB}\zeta^{PDE'} + \chi^D{}_{PAB}\zeta_Q^{PE'} + \overline{\phi}^{E'}{}_{P'AB}\zeta_Q^{DP'}, \qquad (1)$$

and

$$\nabla_{(A'B')}\zeta_Q^{DE'} = \phi_{PQA'B'}\zeta^{PDE'} + \phi^D{}_{PA'B'}\zeta_Q^{PE'} + \overline{\chi}_{P'}{}^{E'}{}_{A'B'}\zeta_Q^{DP'}. \qquad (2)$$

Solution: Equations (1) and (2) are direct generalizations of the results of Problem 6.1 and are left to the reader for verification.

6.3 Prove Eq. (6.193) for the decomposition of the Riemann curvature tensor.

Solution: Equation (6.193) is a straightforward result of the application of Eq. (5.70) and is left to the reader for verification.

6.4 Find the spinor equivalent to the dual of the Riemann curvature tensor.

Solution: The spinor equivalent to the dual of the Riemann curvature tensor is defined by

$${}^\star R_{AB'CD'EF'GH'} = \sigma^\alpha_{AB'}\sigma^\beta_{CD'}\sigma^\gamma_{EF'}\sigma^\delta_{GH'}\, {}^\star R_{\alpha\beta\gamma\delta}, \qquad (1)$$

where ${}^\star R_{\alpha\beta\gamma\delta}$ is the dual to the Riemann curvature tensor given by

$$ {}^\star R_{\alpha\beta\gamma\delta} = \frac{1}{2}\sqrt{-g}R_{\alpha\beta}{}^{\mu\nu}\epsilon_{\mu\nu\gamma\delta}. \qquad (2)$$

The latter formula may also be written in the form

$$ {}^\star R_{\alpha\beta\gamma\delta} = \frac{1}{2}R_{\alpha\beta\mu\nu}\epsilon^{\mu\nu}_{\gamma\delta}, \qquad (3)$$

and therefore in spinor notation in the form

$$ {}^\star R_{AB'CD'EF'GH'} = \frac{1}{2}R_{AB'CD'KL'MN'}\epsilon^{KL'MN'}_{EF'GH'}. \qquad (4)$$

6.5. PROBLEMS

Using now the expression for the spinor equivalent to the Riemann curvature tensor given by Eq. (6.200) and the expression for the spinor $\epsilon_{EF'GH'}^{KL'MN'}$ given by Eq. (5) of Problem 5.3 in the above formula, we then obtain

$$^\star R_{AB'CD'EF'GH'} = i(\chi_{ACEG}\,\epsilon_{B'D'}\,\epsilon_{F'H'} - \phi_{ACF'H'}\,\epsilon_{B'D'}\,\epsilon_{EG}$$

$$+ \epsilon_{AC}\overline{\phi}_{B'D'EG}\,\epsilon_{F'H'} - \epsilon_{AC}\,\epsilon_{EG}\overline{\chi}_{B'D'F'H'}). \tag{5}$$

6.5 Find the spinor equivalent to the tensor

$$C^+_{\alpha\beta\gamma\delta} = C_{\alpha\beta\gamma\delta} + i^\star C_{\alpha\beta\gamma\delta}, \tag{1}$$

where $^\star C_{\alpha\beta\gamma\delta}$ is the dual to the Weyl conformal tensor $C_{\alpha\beta\gamma\delta}$.

Solution: The spinor equivalent to the Weyl conformal tensor is given by Eq. (6.237). The spinor equivalent to the dual of the Weyl tensor may be obtained from that of the Riemann curvature tensor, given by Eq. (5) of Problem 6.4, by replacing the spinor χ_{ABCD} by ψ_{ABCD} and taking $\phi_{ABC'D'} = 0$,

$$^\star C_{AB'CD'EF'GH'} = i\left(\psi_{ACEG}\,\epsilon_{B'D'}\,\epsilon_{F'H'} - \epsilon_{AC}\,\epsilon_{EG}\overline{\psi}_{B'D'F'H'}\right). \tag{2}$$

We consequently obtain

$$C^+_{AB'CD'EF'GH'} = -2\psi_{ACEG}\,\epsilon_{B'D'}\,\epsilon_{F'H'}. \tag{3}$$

6.6 Find the spinor equivalent to the tensor

$$R^+_{\alpha\beta\gamma\delta} = R_{\alpha\beta\gamma\delta} + i\,^\star R_{\alpha\beta\gamma\delta}. \tag{1}$$

Solution: Using Eqs. (6.200) and (6.201) we obtain

$$R^+_{AB'CD'EF'GH'} = -2\left(\chi_{ACEG}\,\epsilon_{B'D'}\,\epsilon_{F'H'} + \epsilon_{AC}\overline{\phi}_{B'D'EG}\,\epsilon_{F'H'}\right). \tag{2}$$

6.7 Find the expression for the spinor $^\star F_{PQAB'CD'}$, the dual to the spinor $F_{PQAB'CD'}$, in terms of the Weyl conformal spinor, the tracefree Ricci spinor, and the Ricci scalar curvature. Show that Eq. (6.190) is identical to the Bianchi identities (6.251).

Solution: The spinor $^\star F_{PQAB'CD'}$ is defined by

$$^\star F_{PQAB'CD'} = \frac{1}{2}\epsilon_{AB'CD'}^{KL'MN'}\,F_{PQKL'MN'}. \tag{1}$$

Using Eq. (2) of Problem 5.3 we then obtain

$$^*F_{PQAB'CD'} = iF_{PQAD'CB'} = i\left(\chi_{PQAC}\,\epsilon_{B'D'} - \phi_{PQB'D'}\,\epsilon_{AC}\right). \qquad (2)$$

Using the above result in Eq. (6.190), we then obtain

$$\nabla^{CD'}\,{}^*F_{PQAB'CD'} = -i\left(\nabla^C{}_{B'}\chi_{PQAC} - \nabla_A{}^{D'}\phi_{PQB'D'}\right) = 0. \qquad (3)$$

Equation (3) is identical to the Bianchi identities (6.251).

6.8 Discuss the physical and geometrical meaning of the field equations

$$\nabla_\nu F_{PQ}{}^{\mu\nu} = 4\pi J_{PQ}{}^\mu, \qquad (1)$$

$$\nabla_\nu\,{}^*F_{PQ}{}^{\mu\nu} = 0, \qquad (2)$$

where $^*F_{PQ}{}^{\mu\nu}$ is the dual to $F_{PQ}{}^{\mu\nu}$ and $J_{PQ}{}^\mu$ represents the energy-momentum tensor, as possible field equations for the theory of gravitation.

Solution: The solution is left for the reader.

6.9 Write the Einstein gravitational field equations in the presence of an electromagnetic field using the spinor calculus.

Solution: The Einstein field equations in the presence of an electromagnetic field have the form $R_{\mu\nu} = (8\pi G/c^4)T_{\mu\nu}$ since $R=0$. The equivalent equations, using spinor calculus, are given by

$$\phi_{ACB'D'} = \frac{4\pi G}{c^4} T_{AB'CD'}, \qquad (1)$$

where $\phi_{ACB'D'}$ is the tracefree Ricci spinor. Using now Eq. (6.217) and Eq. (6) of Problem 5.10, we then obtain

$$\phi_{mn} = \frac{2G}{c^4}\phi_m\bar\phi_n, \qquad (2)$$

for the Einstein field equations. Here $m, n = 0, 1, 2$.

6.10 Prove the transformation laws (6.15) and (6.16) of the Christoffel symbols of the first and second kinds.

Solution: Using the transformation laws for the metric tensor leads to Eqs. (6.15) and (6.16).

6.11 Prove Eq. (6.24).

6.5. PROBLEMS

Solution: The solution is left for the reader.

6.12 Show that the covariant derivatives of the tensors $T_{\alpha\beta}$, $T^{\alpha\beta}$ and $T^{\alpha}{}_{\beta}$ are given by

$$\nabla_\gamma T_{\alpha\beta} = \frac{\partial T_{\alpha\beta}}{\partial x^\gamma} - \Gamma^\delta_{\beta\gamma} T_{\alpha\delta} - \Gamma^\delta_{\alpha\gamma} T_{\delta\beta},$$

$$\nabla_\gamma T^{\alpha\beta} = \frac{\partial T^{\alpha\beta}}{\partial x^\gamma} + \Gamma^\alpha_{\delta\gamma} T^{\delta\beta} + \Gamma^\beta_{\delta\gamma} T^{\alpha\delta},$$

$$\nabla_\gamma T^\alpha{}_\beta = \frac{\partial T^\alpha{}_\beta}{\partial x^\gamma} + \Gamma^\alpha_{\delta\gamma} T^\delta{}_\beta - \Gamma^\delta_{\beta\gamma} T^\alpha{}_\delta.$$

From this find the general rule for covariant differentiation.

Solution: The solution is left for the reader.

6.13 Show that the covariant differentiation of the sum, difference, outer and inner products of tensors obeys the usual rules of ordinary differentiation.

Solution: The solution is left for the reader.

6.14 Generalize Eq. (6.30) for a tensor $T_{\mu\nu}$.

Solution: The solution is left for the reader.

6.15 If $T_{\alpha\beta}$ is the curl of a covariant vector, show that

$$\nabla_\gamma T_{\alpha\beta} + \nabla_\alpha T_{\beta\gamma} + \nabla_\beta T_{\gamma\alpha} = 0,$$

and that this is equivalent to

$$\partial_\gamma T_{\alpha\beta} + \partial_\alpha T_{\beta\gamma} + \partial_\beta T_{\gamma\alpha} = 0.$$

Solution: The solution is left for the reader.

6.16 Show that the divergence $\nabla_\mu V^\mu$ of the vector V^μ is given by

$$\nabla_\mu V^\mu = \frac{1}{\sqrt{-g}} \frac{\partial}{\partial x^\mu} \left(V^\mu \sqrt{-g}\right).$$

Also show that for a skew-symmetric tensor $F^{\alpha\beta}$ the covariant divergence is

$$\nabla_\beta F^{\alpha\beta} = \frac{1}{\sqrt{-g}} \frac{\partial}{\partial x^\beta} \left(F^{\alpha\beta} \sqrt{-g}\right).$$

Solution: The solution is left for the reader.

6.17 Find the expression for the Riemann tensor $R_{\alpha\beta\gamma\delta}$. From it prove Eqs. (6.32).

Solution: The solution is left for the reader.

6.18 Show that a curve with a covariantly constant tangent vector is necessarily geodesic.

Solution: Let the curve be denoted by $x^\alpha = x^\alpha(s)$ and the tangent vector by dx^α/ds. If the tangent vector is covariantly constant, then

$$\nabla_\mu \frac{dx^\alpha}{ds} = 0, \tag{1}$$

or explicitly

$$\frac{\partial}{\partial x^\mu}\left(\frac{dx^\alpha}{ds}\right) + \Gamma^\alpha_{\mu\nu}\frac{dx^\nu}{ds} = 0. \tag{2}$$

Multiplying Eq. (2) by dx^μ/ds and using the identity

$$\frac{dx^\mu}{ds}\frac{\partial}{\partial x^\mu} = \frac{d}{ds}, \tag{3}$$

gives

$$\frac{d^2 x^\alpha}{ds^2} + \Gamma^\alpha_{\mu\nu}\frac{dx^\mu}{ds}\frac{dx^\nu}{ds} = 0. \tag{4}$$

6.19 Discuss the constancy of the weak and gravitational coupling constants.

Solution: The solution is left for the reader.

6.20 Use the geodesic equations, Eq. (6.43), to determine the force per unit mass on a body at rest, and show that it is given by $F^i = -c^2\Gamma^i_{00}$ where $i = 1, 2, 3$. In the weak field approximation $g^{i\alpha}$ are very close to the Lorentz metric, and for a time-independent metric $F^i = c^2\Gamma^i_{00} \approx (c^2/2)\partial_i g_{00}$. Show that in the weak field case Eq. (6.50) reduces to the Poisson equation (6.49) where $g_{00} \approx 1 + 2\phi/c^2$. From this show that the constant κ in Eq. (6.50) is given by $\kappa = 8\pi G/c^4$.

Solution: The solution is left for the reader.

6.21 Prove Eqs. (6.58) and (6.59).

Solution: The solution is left for the reader.

6.22 Derive the gravitational field equations (6.50) using the calculus of variation by treating both $g_{\mu\nu}$ and $\Gamma^\mu_{\alpha\beta}$ as independent variants, and obtain thereby equations that determine both objects. Such a procedure is known as the Palatini formalism. The procedure is analogous to the one employed in deriving the electromagnetic field equations from a variational principle where both the field $f^{\mu\nu}$ and the potential A_μ are variants of an action principle.

Solution: The solution is left for the reader.

6.23 Find the energy-momentum tensor $T_{\mu\nu}$ for: (1) a system of neutral particles of inertial mass M (function of time); (2) the electromagnetic field; and (3) a scalar field ϕ. Show that they are given by:

$$(1)\ T^{\mu\nu} = \sum M\dot{z}^\mu \dot{z}^\nu \delta(\mathbf{x} - \mathbf{z}),$$

$$(2)\ T_{\mu\nu} = \frac{1}{4\pi}\left\{\frac{1}{4}g_{\mu\nu}f_{\alpha\beta}f^{\alpha\beta} - f_{\mu\alpha}f_\nu{}^\alpha\right\},$$

$$(3)\ T_{\mu\nu} = \partial_\mu\phi\partial_\nu\phi - \frac{1}{2}g_{\mu\nu}\left(\nabla^\alpha\phi\nabla_\alpha\phi - m^2\phi^2\right).$$

Solution: The solution is left for the reader.

6.24 Use the approximate metric (6.79) into the geodesic equation (6.43) to show that the equation obtained is (6.80).

Solution: The solution is left for the reader.

6.25 Prove Eqs. (6.150) and (6.151).

Solution: The solution is left for the reader.

6.6 References for Further Reading

W.L. Bade and J. Jehle, An introduction to spinors, *Revs. Mod. Phys.* **25**, 714-728 (1953). (Sections 6.2-6.4)

B. Bertotti, D. Brill and R. Krotkov, Experiments on gravitation, in: *Gravitation: An Introduction to Current Research* (L. Witten, Editor), (John Wiley, New York, 1962). (Section 6.1)

G. Birkhoff, *Relativity and Modern Physics* (Harvard University Press, Cambridge, Mass., 1923). (Section 6.1)

M. Carmeli, *Group Theory and General Relativity* (McGraw-Hill, New York, 1977).

M. Carmeli, *Classical Fields: General Relativity and Gauge Theory* (John

Wiley, New York, 1982).

M. Carmeli, Equations of motion without infinite self-action terms in general relativity, *Phys. Rev.* **140**, B1441 (1965). (Section 6.1)

M. Carmeli, E. Leibowitz and N. Nissani, *Gravitation: SL(2,C) Gauge Theory and Conservation Laws* (World Scientific, Singapore, 1990). (Sections 6.2-6.4)

T.F. Cranshaw, S.P. Schiffer and A.B. Whitehead, *Phys. Rev. Letters* **4**, 163 (1960). (Section 6.1)

R.H. Dicke, in: *Relativity, Groups and Topology* (C. DeWitt *et al.*, Eds.) (Gordon and Breach, New York, 1964). (Section 6.1)

A. Einstein, *Ann. Phys.* **49**, 761 (1916); English translation in: *The Principle of Relativity* (Dover, New York, 1923). (Section 6.1)

L.P. Eisenhart, *Riemannian Geometry* (Princeton University Press, New Jersey, 1949). (Section 6.1)

L. Infeld and A. Schild, *Rev. Mod. Phys* **21**, 408 (1949). (Section 6.1)

L. Infeld and B.L. van der Waerden, The wave equation of the electron in the general relativity theory, *Sb. preuss. Akad. Wiss., Phys.-mat.* Kl. **380** (1933). (Sections 6.2-6.4)

A. Papapetrou, *Proc. R. Soc. London (A)* **209**, 248 (1951). (Section 6.1)

R. Penrose, A spinor approach to general relativity, *Ann. Phys. (N.Y.)* **10**, 171-201 (1960). (Sections 6.2-6.4)

F.A.E. Pirani, Introduction to gravitational radiation theory, in: *Lectures on General Relativity* (1964 Brandeis Summer School) (Prentice-Hall, Englewood Cliffs, New Jersey, 1965). (Section 6.1)

R.V. Pound and G.A. Rabka, Jr., *Phys. Rev. Letters* **4**, 337 (1960). (Section 6.1)

P.G. Roll, R. Krotkov and R.H. Dicke, *Ann. Phys. (N.Y.)* **26**, 442 (1964). (Section 6.1)

L.I. Schiff, *Proc. Natl. Acad. Sci.* **46**, 871 (1960). (Section 6.1)

I.I. Shapiro, *General Relativity and Gravitation* **3**, 135 (1972). (Section 6.1)

A. Trautman, Foundations and current problems of general relativity, in: *Lectures on General Relativity* (1964 Brandeis Summer School) (Prentice-

6.6. REFERENCES FOR FURTHER READING

Hall, Englewood Cliffs, New Jersey, 1965). (Section 6.1)

O. Veblen and J. von Neumann, *Geometry of Complex Dynamics* (The Institute for Advanced Study, Princeton, New Jersey, 1958). (Sections 6.2-6.4)

J. Weber, Gravitational radiation experiments, in: *Relativity* (M. Carmeli, S.I. Fickler and L. Witten, Eds.), (Plenum Press, New York, 1970). (Section 6.1)

S. Weinberg, *Gravitation and Cosmology* (John Wiley, New York, 1973). (Section 6.1)

Chapter 7

The Gauge Field Spinors

After discussing the electromagnetic and the gravitational fields in the last two chapters, we now formulate the gauge field dynamical variables in a spinorial form. Following a brief review of the Yang-Mills theory, the spinors equivalent to the gauge potential and the gauge field strength are written down. Then, in analogy to the electromagnetic field, the gauge field strength spinor is decomposed. Likewise, the expression for the energy-momentum tensor of the gauge field is given in a spinorial form. This is subsequently followed by formulating the gauge field variables as spinors in the interior spaces of both the groups SL(2,C) and SU(2). The chapter is then concluded with formulating and analyzing the geometry of the gauge field dynamical variables.

7.1 The Yang-Mills Theory

In this section a brief review of the Yang-Mills theory is given.

7.1.1 Gauge Invariance

In ordinary gauge invariance of a charged field which is described by a complex wave function ψ, a change of *gauge* means a change of *phase factor* $\psi \to \psi'$, $\psi' = (\exp i\alpha)\psi$, a change that is devoid of any physical consequences. Since ψ depends on spacetime points, the relative phase factor of ψ at two different points is completely arbitrary and α is, accordingly, a function of spacetime. In other words, the arbitrariness in choosing the phase factor is local in character.

To preserve invariance it is then necessary to counteract the variation of the phase α with spacetime coordinates by introducing the electromagnetic potentials $A_\mu(x)$ which change under a gauge transformation as

$$A'_\mu = A_\mu + \frac{1}{e}\frac{\partial \alpha}{\partial x^\mu}, \qquad (7.1)$$

and to replace the derivative of ψ by a "covariant derivative" with the combination $(\partial_\mu - ieA_\mu)\psi$.

7.1.2 Isotopic Spin

An *isotopic spin* parameter was first introduced by Heisenberg in 1932 to describe the two charge states, namely neutron and proton, of a nucleon. The idea that the neutron and proton correspond to two states of the same particle was suggested at the same time by the fact that their masses are nearly equal, and the light stable even nuclei contain equal numbers of them.

Later on it was pointed out that the $p-p$ and $n-p$ interactions are approximately equal in the ^1S state, and consequently it was assumed that the equality holds also in the other states available to both the $n-p$ and $p-p$ systems.

Under such an assumption one arrives at the concept of a *total* isotopic spin which is conserved in nucleon-nucleon interactions. Experiments on the energy levels of light nuclei strongly suggest that this assumption is indeed correct. This implies that all strong interactions, such as the pion-nucleon interaction, should also satisfy the same conservation law. This, and the fact that there are three charge states for the pion can be coupled to the nucleon field singly, lead to the conclusion that pions have isotopic spin unity. A verification of this conclusion was found in experiments which compare the differential cross-section of the process $n + p \to \pi^0 + d$ with that of the perviously measured process $p + p \to \pi^+ + d$.

7.1.3 Conservation of Isotopic Spin and Invariance

The conservation of isotopic spin is identical with the requirement that all interactions be invariant under isotopic spin rotation, when electromagnetic interactions are neglected. This means that the orientation of the isotopic spin has no physical significance.

Differentiation between a neutron and a proton is then an arbitrary process. This arbitrariness is subject to the limitation that once one chooses

7.1. THE YANG-MILLS THEORY

what to call a proton and what to call a neutron at one spacetime point, one is then not free to make any other choices at other spacetime points. It also seems not to be consistent with the localized field concept which underlies the usual physical theories.

7.1.4 Isotopic Spin and Gauge Fields

The possibility of requiring that all interactions to be invariant under *independent* rotations of the isotopic spin at all spacetime points, so that the relative orientation of the isotopic spin at two spacetime points becomes physically meaningless, was accordingly explored by Yang and Mills. They introduced *isotopic gauge* as an arbitrary way of choosing the orientation of the isotopic spin axes at all spacetime points, in analogy with the electromagnetic gauge which represents an arbitrary way of choosing the complex phase factor of a charged field at all spacetime points.

This suggests that all physical processes, which do not involve the electromagnetic field, be invariant under the isotopic gauge transformation $\psi \to \psi'$, $\psi' = S^{-1}\psi$, where S represents a spacetime dependent isotopic spin rotation which is a 2×2 unitary matrix with determinant unity, i.e., an element of the group SU(2) discussed in Chapter 1.

In an entirely similar manner to what is done in electromagnetics, Yang and Mills introduced a potential B in the case of the isotopic gauge transformation to counteract the dependence of the matrix S on the spacetime coordinates.

Accordingly, and in analogy with the electromagnetic case, all derivatives of the wave function ψ describing a field with isotopic spin $\frac{1}{2}$ should appear as "covariant derivatives" of the form $(\partial_\mu - iB_\mu)\psi$, where B_μ are four 2×2 Hermitian matrices. The field equations satisfied by the twelve independent components of the B potential, which is called the **b** potential, and their interaction with any field having an isotopic spin, are fixed just as in the electromagnetic case.

7.1.5 Isotopic Gauge Transformation

Under an isotopic gauge transformation, a two-component wave function ψ describing a field with isotopic spin $\frac{1}{2}$, transforms according to

$$\psi = S\psi'. \tag{7.2}$$

Invariance then requires that the covariant derivative expression transforms as

$$S\left(\partial_\mu - iB'_\mu\right)\psi' = \left(\partial_\mu - iB_\mu\right)\psi. \tag{7.3}$$

When combined with Eq. (7.2), we obtain the isotopic gauge transformation of the 2×2 potential matrix B_μ:

$$B'_\mu = S^{-1} B_\mu S + i S^{-1} \partial_\mu S. \tag{7.4}$$

In analogy to the procedure of obtaining gauge invariant fields in the electromagnetic case, Yang and Mills defined their field as

$$F_{\mu\nu} = \partial_\nu B_\mu - \partial_\mu B_\nu + i [B_\mu, B_\nu], \tag{7.5}$$

where the commutator is given by

$$[B_\mu, B_\nu] = B_\mu B_\nu - B_\nu B_\mu. \tag{7.6}$$

Under the transformation (7.2) the 2×2 field matrix (7.5) transforms as

$$F'_{\mu\nu} = S^{-1} F_{\mu\nu} S. \tag{7.7}$$

Now Eq. (7.4) is valid for any S and its corresponding B_μ. Furthermore, the matrix $S^{-1} \partial S / \partial x^\mu$ appearing in Eq. (7.4) is a linear combination of the isotopic spin "angular momentum" matrices T^i, $i = 1, 2, 3$, corresponding to the isotopic spin on the field ψ under consideration. Here $T^i = \frac{1}{2}\sigma^i$, where σ^i are the three Pauli spin matrices.

Accordingly, the matrix B_μ itself must also contain a linear combination of the matrices T^i; any part of B_μ in addition to this, denote it by $\tilde{B}_\mu$, is a scalar or tensor combination of the T's, and must transform by the homogeneous part of (7.4),

$$\tilde{B}'_\mu = S^{-1} \tilde{B}_\mu S. \tag{7.8}$$

Such a field is extraneous and was allowed by the very general form we took for the B potential, but is irrelevant to the question of isotopic gauge. Therefore, the relevant part of the B potential can be written as a linear combination of the matrices T^i:

$$B_\mu = 2\mathbf{b}_\mu \cdot \mathbf{T}, \tag{7.9}$$

where bold-face letters denote 3-component vectors in the isospin space.

The isospin-gauge covariant field matrices $F_{\mu\nu}$ can also be expressed as a linear combination of the T's. One obtains

$$F_{\mu\nu} = 2\mathbf{f}_{\mu\nu} \cdot \mathbf{T}, \tag{7.10}$$

7.1. THE YANG-MILLS THEORY

where

$$f_{\mu\nu} = \frac{\partial \mathbf{b}_\mu}{\partial x^\nu} - \frac{\partial \mathbf{b}_\nu}{\partial x^\mu} + 2\mathbf{b}_\mu \times \mathbf{b}_\nu. \tag{7.11}$$

One notices that $\mathbf{f}_{\mu\nu}$ transforms like a vector under an isotopic gauge transformation. The corresponding transformation of $\mathbf{b}_\mu$ is cumbersome. Under *infinitesimal* isotopic gauge transformations,

$$S = 1 - 2i\mathbf{T} \cdot \delta\mathbf{w}, \tag{7.12}$$

however, one obtains

$$\mathbf{b}'_\mu = \mathbf{b}_\mu - 2\mathbf{b}_\mu \times \delta\mathbf{w} + \frac{\partial \delta \mathbf{w}}{\partial x^\mu}. \tag{7.13}$$

7.1.6 Field Equations

In analogy to the electromagnetic case one can write down an isotopic gauge invariant Lagrangian density:

$$L = -\frac{1}{8}\mathrm{Tr} F_{\mu\nu} F^{\mu\nu} = -\frac{1}{4} \mathbf{f}_{\mu\nu} \cdot \mathbf{f}^{\mu\nu}. \tag{7.14}$$

One can also include a field with isotopic spin $\frac{1}{2}$ to obtain the following total Lagrangian density:

$$L = -\frac{1}{8}\mathrm{Tr} F_{\mu\nu} F^{\mu\nu} - \overline{\psi}\gamma_\mu \left(\partial_\mu - iB_\mu\right)\psi - m\overline{\psi}\psi. \tag{7.15}$$

The equations of motion obtained from the Lagrangian (7.15) are

$$\frac{\partial \mathbf{f}^{\mu\nu}}{\partial x^\nu} - 2\left(\mathbf{b}_\nu \times \mathbf{f}^{\mu\nu}\right) + \mathbf{J}^\mu = 0, \tag{7.16}$$

$$\gamma^\mu \left(\partial_\mu - 2i\mathbf{T} \cdot \mathbf{b}_\mu\right)\psi + m\psi = 0, \tag{7.17}$$

where

$$\mathbf{J}_\mu = 2i\overline{\psi}\gamma_\mu \mathbf{T}\psi. \tag{7.18}$$

Continuity equation

Since the divergence of $\mathbf{J}^\mu$ does not vanish, one may define

$$\tilde{\mathbf{J}}^\mu = \mathbf{J}^\mu - 2\mathbf{b}_\nu \times \mathbf{f}^{\mu\nu}, \tag{7.19}$$

which leads to the *equation of continuity*,

$$\frac{\partial \tilde{\mathbf{J}}^\mu}{\partial x^\mu} = 0. \tag{7.20}$$

Equation (7.20) guarantees that the total isotopic spin

$$T = \int \tilde{\mathbf{J}}^0 d^3 x, \tag{7.21}$$

is independent of time and Lorentz transformation.

Nonlinearity of the Field Equations

Equation (7.19) shows that the isotopic spin arises from both the spin-$\frac{1}{2}$ field $\mathbf{J}^\mu$ and from the $\mathbf{b}_\mu$ potential itself. This fact makes the field equations for the B potential *nonlinear*, even in the absence of the spin-$\frac{1}{2}$ field. The situation here is different from that of the electromagnetic case whose field is chargeless, and hence satisfies linear equations.

7.2 Gauge Potential and Field Strength

The spinors equivalent to SU(2) gauge potential $b_{a\mu}$ and gauge field strength $f_{a\mu\nu}$ are complex functions, which are obtained from the potential and the field strength in the same way that the comparable spinors are obtained in the theory of electrodynamics (see Section 5.4). The gauge potential and field strength are related by the equation

$$f_{a\mu\nu} = \partial_\nu b_{a\mu} - \partial_\mu b_{a\nu} + g\epsilon_{abc} b_{b\mu} b_{c\nu}, \tag{7.22}$$

where g is a coupling constant and ϵ_{abc} is the skew-symmetric tensor defined by $\epsilon_{123} = 1$. In the above quantities the indices $a, b, c = 1, 2, 3$ are SU(2) labels describing the inner space degrees of freedom, whereas $\mu, \nu = 0, 1, 2, 3$, are spacetime indices.

7.2.1 The Yang-Mills Spinor

The spinor equivalent to the gauge potential is given by

$$b_{aAB'} = \sigma^\mu_{AB'} b_{a\mu}, \tag{7.23}$$

whereas that equivalent to the gauge field strength is given by

$$f_{aAB'CD'} = \sigma^\mu_{AB'} \sigma^\nu_{CD'} f_{a\mu\nu}. \tag{7.24}$$

7.2. GAUGE POTENTIAL AND FIELD STRENGTH

Since the potential $b_{a\mu}$ is real, its spinor equivalent $b_{aAB'}$ is Hermitian, namely,
$$b_{aAB'} = \overline{b}_{aB'A}. \tag{7.25}$$
Accordingly $b_{a00'}$ and $b_{a11'}$ are real quantities, whereas $b_{a01'}$ and $b_{a10'}$ are complex quantities, conjugate to each other,
$$b_{a10'} = \overline{b}_{a0'1} = \overline{b_{a01'}}. \tag{7.26}$$

Decomposition of the Spinor Equivalent

In analogy to the decomposition given by Eq. (5.83) of the spinor equivalent to the electromagnetic field tensor, the spinor equivalent to the gauge field strength may be decomposed. We then obtain
$$f_{aAB'CD'} = \chi_{aAC}\,\epsilon_{B'D'} + \epsilon_{AC}\overline{\chi}_{aB'D'}, \tag{7.27}$$
where
$$\chi_{aAC} = \chi_{aCA} = \frac{1}{2}\epsilon^{B'D'} f_{aAB'CD'}. \tag{7.28}$$
Since the spinor χ_{aAB} is symmetric in its spinor indices A and B, it has 3×3 complex components: χ_{a00}, $\chi_{a01} = \chi_{a10}$, and χ_{a11}, with $a = 1, 2, 3$. These nine complex components are equivalent to the 18 real components of the field strength $f_{a\mu\nu}$.

The Yang-Mills Spinor

The gauge field spinor χ_{aAB} will be referred to in the sequel as the *Yang-Mills spinor*. Its role is analogous to the Maxwell spinor ϕ_{AB} in electromagnetics (see Section 5.4). We will also, sometimes, use the notation

$$\chi_{a0} = \chi_{a00}$$
$$\chi_{a1} = \chi_{a01} = \chi_{a10} \tag{7.29}$$
$$\chi_{a2} = \chi_{a11}$$

in analogy to the Maxwell spinor.

We may also find the spinor equivalent to the tensor ${}^\star f_{a\mu\nu}$, where
$${}^\star f_{a\mu\nu} = \frac{1}{2}\sqrt{-g}\,\epsilon_{\mu\nu\rho\sigma}\,f_a{}^{\rho\sigma}, \tag{7.30}$$

which is the dual to the gauge field strength $f_{a\mu\nu}$. We then find

$$^\star f_{aAB'CD'} = \frac{1}{2}\epsilon_{AB'CD'}^{KL'MN'} f_{aKL'MN'}, \qquad (7.31)$$

or, using Eq. (2) of Problem 5.3,

$$^\star f_{aAB'CD'} = if_{aAD'CB'} = i\left(\epsilon_{AC}\overline{\chi}_{aB'D'} - \chi_{aAC}\,\epsilon_{B'D'}\right). \qquad (7.32)$$

The spinor equivalent to the tensor

$$f_{a\rho\sigma}^+ = f_{a\rho\sigma} + i\,{}^\star f_{a\rho\sigma} \qquad (7.33)$$

is consequently given by

$$f_{aAB'CD'}^+ = f_{aAB'CD'} - f_{aAD'CB'} = 2\chi_{aAC}\epsilon_{B'D'}, \qquad (7.34)$$

where use has been made of Eq. (7.27).

Likewise, we may find the spinor equivalent to the tensor

$$f_{a\rho\sigma}^- = f_{a\rho\sigma} - i\,{}^\star f_{a\rho\sigma}. \qquad (7.35)$$

We then obtain

$$f_{aAB'CD'}^- = f_{aAB'CD'} + f_{aAD'CB'} = 2\epsilon_{AC}\overline{\chi}_{aB'D'}. \qquad (7.36)$$

Accordingly we have

$$f_{aAB'CD'} = \frac{1}{2}\left(f_{aAB'CD'}^+ + f_{aAB'CD'}^-\right), \qquad (7.37)$$

and

$$^\star f_{aAB'CD'}^{\mp} = \pm i f_{aAB'CD'}^{\mp}, \qquad (7.38)$$

for the duals of $f_{aAB'CD'}^{\mp}$.

7.2.2 Energy-Momentum Spinor

The energy-momentum tensor of a gauge field is given by

$$T_{\mu\nu} = \frac{1}{4\pi}\left(\frac{1}{4}g_{\mu\nu}f_{a\alpha\beta}f_a{}^{\alpha\beta} - f_{a\mu\alpha}f_{a\nu}{}^\alpha\right), \qquad (7.39)$$

and is in complete analogy to that of the electromagnetic field. It may also be written in the form

$$T_{\mu\nu} = -\frac{1}{8\pi}\left(f_{a\mu\alpha}f_{a\nu}{}^\alpha + {}^\star f_{a\mu\alpha}^\star f_{a\nu}{}^\alpha\right), \qquad (7.40)$$

7.2. GAUGE POTENTIAL AND FIELD STRENGTH

and is, as can easily be seen, traceless,

$$T_\mu{}^\mu = 0, \tag{7.41}$$

just as the case in electrodynamics.

The spinor equivalent to the energy-momentum tensor of the gauge field is then given by

$$T_{AB'CD'} = -\frac{1}{8\pi}\left(f_{aAB'EF'}f_{aCD'}{}^{EF'} + {}^*f_{aAB'EF'}{}^*f_{aCD'}{}^{EF'}\right). \tag{7.42}$$

Using now the expressions for f and $*f$ given by Eqs. (7.27) and (7.32) in Eq. (7.42), we then obtain

$$T_{AB'CD'} = \frac{1}{2\pi}\chi_{aAC}\overline{\chi}_{aB'D'}. \tag{7.43}$$

If we denote the above spinor by

$$T_{mn} = T_{A+C,B'+D'}, \tag{7.44}$$

with $m, n = 0, 1, 2$, Eq. (7.43) will then have the form

$$T_{mn} = \frac{1}{2\pi}\chi_{am}\overline{\chi}_{an}. \tag{7.45}$$

This form for the energy-momentum tensor is in complete analogy to that of the electromagnetic field (see Eq. (6) of Problem 5.10).

The Einstein Field Equations

The Einstein field equations with the above energy-momentum tensor have the form

$$R_{\mu\nu} = \frac{8\pi G}{c^4}T_{\mu\nu}, \tag{7.46}$$

since the Ricci scalar curvature $R = -\left(8\pi G/c^4\right)T_\mu{}^\mu = 0$ by Eq. (7.41). The equivalent gravitational field equations, using the spinor notation, are given by

$$\phi_{ACB'D'} = \frac{4\pi G}{c^4}T_{AB'CD'}, \tag{7.47}$$

where use has been made of Eqs. (6.222) and (6.223). Using now Eqs. (6.217) and (7.40), we then obtain for the Einstein field equations in the presence of a gauge field the following:

$$\phi_{mn} = \frac{2G}{c^4}\chi_{am}\overline{\chi}_{an}. \tag{7.48}$$

Here $m, n = 0, 1, 2$, and $a = 1, 2, 3$. We notice that Eq. (7.48) is in complete analogy to Eq. (2) of Problem 6.9 for the case of the Einstein equations in the presence of an electromagnetic field.

7.2.3 SU(2) Spinors

So far we have described the gauge field and potential in terms of SL(2,C) spinors, leaving the SU(2) inner space degree of freedom indices unchanged. We now develop an SU(2) spinor calculus to take care of that degree of freedom. The Yang-Mills spinor χ_{aAB}, for instance, will thus be described as a quantity having two SL(2,C) spinor indices and two SU(2) spinor indices, χ_{MNAB}, where $M, N = 0, 1$ also.

The relationship between isospinors and isovectors is as follows. An isospin-$\frac{1}{2}$ object is described by an SU(2) spinor with one index. An example of this is the proton and the neutron which are described collectively by the spinor ψ_M (it is, in addition, a four-component Dirac spinor in configuration space). An isospin-1 object is described by a two-index SU(2) spinor which is symmetric in the two indices. An isospin-T object is described by a totally symmetric spinor having $2T$ indices. It therefore has $2T+1$ independent components.

The correspondence between isovectors and isospinors is achieved by means of the Pauli spin matrices. The spinor equivalent to the vector ξ_a is given by

$$\xi_M{}^N = \sigma_{aM}{}^N \xi_a, \qquad (7.49)$$

whereas the isovector equivalent to the spinor $\xi_M{}^N$ is given by

$$\xi_a = \sigma_{aM}{}^N \xi_N{}^M. \qquad (7.50)$$

Here $\sigma_{aM}{}^N$ are the usual three Pauli matrices divided by $\sqrt{2}$:

$$\sigma_{1M}{}^N = \frac{1}{\sqrt{2}}\begin{pmatrix} 0 & 1 \\ 1 & 0 \end{pmatrix}, \quad \sigma_{2M}{}^N = \frac{1}{\sqrt{2}}\begin{pmatrix} 0 & i \\ -i & 0 \end{pmatrix},$$

$$\sigma_{3M}{}^N = \frac{1}{\sqrt{2}}\begin{pmatrix} 1 & 0 \\ 0 & -1 \end{pmatrix}. \qquad (7.51)$$

7.2.4 Spinor Indices

The SU(2) spinor indices for the Pauli matrices are chosen in such a way that the spinor equivalent to the isovector is symmetric when both indices are upper or lower:

$$\xi_{MN} = \xi_M{}^P \epsilon_{PN} = \sigma_{aM}{}^P \epsilon_{PN} \xi_a = \sigma_{aMN} \xi_a = \xi_{NM}, \qquad (7.52)$$

7.3. THE GEOMETRY OF GAUGE FIELDS

$$\xi^{MN} = \epsilon^{MP}\xi_P{}^N = \epsilon^{MP}\sigma_{aP}{}^N\xi_a = \sigma_a{}^{MN}\xi_a = \xi^{NM}. \quad (7.53)$$

Equations (7.52) and (7.53) are the consequence of the symmetry of the matrices σ_{aMN} and σ_a^{MN}. In fact we have

$$\sigma_{1MN} = \frac{1}{\sqrt{2}}\begin{pmatrix} -1 & 0 \\ 0 & 1 \end{pmatrix}, \quad \sigma_{2MN} = \frac{1}{\sqrt{2}}\begin{pmatrix} -i & 0 \\ 0 & -i \end{pmatrix},$$

$$\sigma_{3MN} = \frac{1}{\sqrt{2}}\begin{pmatrix} 0 & 1 \\ 1 & 0 \end{pmatrix}, \quad (7.54)$$

and

$$\sigma_1^{MN} = \frac{1}{\sqrt{2}}\begin{pmatrix} 1 & 0 \\ 0 & -1 \end{pmatrix}, \quad \sigma_2^{MN} = \frac{1}{\sqrt{2}}\begin{pmatrix} -i & 0 \\ 0 & -i \end{pmatrix},$$

$$\sigma_3^{MN} = \frac{1}{\sqrt{2}}\begin{pmatrix} 0 & -1 \\ -1 & 0 \end{pmatrix}, \quad (7.55)$$

as compared to our previous presentation for the Pauli matrices in the SL(2,C) spinor calculus given by Eqs. (5.46).

The Yang-Mills spinor χ_{aAB}, for instance, will be presented by the mixed SU(2) and SL(2,C) spinor χ_{MNAB}. It is symmetric with its two kinds of indices, namely,

$$\chi_{MNAB} = \chi_{NMAB}, \quad \chi_{MNAB} = \chi_{MNBA}. \quad (7.56)$$

We also notice that in the SU(2) spinor calculus there are no primed indices, and that we raise and lower the indices with ϵ^{MN} and ϵ_{MN} just as for the SL(2,C) spinor case.

In the next section we give the geometry of the Yang-Mills fields.

7.3 The Geometry of Gauge Fields

We now give to the gauge field dynamical variables a geometrical description.

7.3.1 Four-Index Tensor

Let $f_{a\mu\nu}$ be the gauge field strengths. Here $\mu, \nu = 0, 1, 2, 3$ are the spacetime indices, and a is the isospin index taking the values 1, 2, 3. From the field strengths we define the four-index tensor

$$R_{\mu\nu\rho\sigma} = -f_{a\mu\nu}f_{a\rho\sigma}. \quad (7.57)$$

The tensor $R_{\mu\nu\rho\sigma}$, which is an SU(2) invariant, satisfies the symmetry properties

$$R_{\mu\nu\rho\sigma} = -R_{\nu\mu\rho\sigma} = -R_{\mu\nu\sigma\rho} = +R_{\rho\sigma\mu\nu}. \tag{7.58}$$

Hence the tensor $R_{\mu\nu\rho\sigma}$ is skew-symmetric in each pair of the indices $\mu\nu$ and $\rho\sigma$, and is symmetric under the exchange of these two pairs of indices with each other. These symmetry properties are the same as those of the Riemann curvature tensor, or the Weyl conformal tensor, known from the geometry of curved spacetime (see the book of Eisenhart).

It will also be useful to define another tensor $R^\star_{\mu\nu\rho\sigma}$, which is also an SU(2) gauge invariant, by

$$R^\star_{\mu\nu\rho\sigma} = -f_{a\mu\nu}\,{}^\star f_{a\rho\sigma}. \tag{7.59}$$

Here $^\star f_{a\rho\sigma}$ is the dual to the tensor $f_{a\rho\sigma}$,

$$^\star f_{a\rho\sigma} = \frac{1}{2}\sqrt{-g}\epsilon_{\rho\sigma\mu\nu} f_a^{\mu\nu}. \tag{7.60}$$

The tensor $R^\star_{\mu\nu\rho\sigma}$ has the same symmetry properties as those of $R_{\mu\nu\rho\sigma}$, namely

$$R^\star_{\mu\nu\rho\sigma} = -R^\star_{\nu\mu\rho\sigma} = -R^\star_{\mu\nu\sigma\rho} = +R^\star_{\rho\sigma\mu\nu}. \tag{7.61}$$

From the two tensors $R_{\mu\nu\rho\sigma}$ and $R^\star_{\mu\nu\rho\sigma}$ we may then define the complex tensor

$$\tilde{R}_{\mu\nu\rho\sigma} = R_{\mu\nu\rho\sigma} + iR^\star_{\mu\nu\rho\sigma} = -f_{a\mu\nu} f^+_{a\rho\sigma}, \tag{7.62}$$

where

$$f^+_{a\rho\sigma} = f_{a\rho\sigma} + i\,{}^\star f_{a\rho\sigma}. \tag{7.63}$$

The new tensor $\tilde{R}_{\mu\nu\rho\sigma}$ also satisfies the same symmetry properties of $R_{\mu\nu\rho\sigma}$ and $R^\star_{\mu\nu\rho\sigma}$,

$$\tilde{R}_{\mu\nu\rho\sigma} = -\tilde{R}_{\nu\mu\rho\sigma} = -\tilde{R}_{\mu\nu\sigma\rho} = +\tilde{R}_{\rho\sigma\mu\nu}. \tag{7.64}$$

From the tensor $\tilde{R}_{\mu\nu\rho\sigma}$ we may define the Ricci tensor $\tilde{R}_{\alpha\beta} = \tilde{R}^\sigma{}_{\alpha\sigma\beta}$ and the Ricci scalar curvature $\tilde{R} = \tilde{R}^\alpha{}_\alpha$.

Since the tensor $\tilde{R}_{\alpha\beta\gamma\delta}$ has the same symmetry properties (except for the cyclic identity) as those of the Riemann curvature tensor, we may decompose it as follows:

$$\tilde{R}_{\rho\sigma\mu\nu} = \tilde{C}_{\rho\sigma\mu\nu} + \frac{1}{2}\left(g_{\rho\mu}\tilde{R}_{\sigma\nu} - g_{\rho\nu}\tilde{R}_{\sigma\mu} - g_{\sigma\mu}\tilde{R}_{\rho\nu} + g_{\sigma\nu}\tilde{R}_{\rho\mu}\right)$$

$$+ \frac{1}{6}\left(g_{\rho\nu}g_{\sigma\mu} - g_{\rho\mu}g_{\sigma\nu}\right)\tilde{R}, \tag{7.65}$$

7.3. THE GEOMETRY OF GAUGE FIELDS

or in the alternative, but equivalent, form

$$\tilde{R}_{\rho\sigma\mu\nu} = \tilde{C}_{\rho\sigma\mu\nu} + \frac{1}{2}\left(g_{\rho\mu}\tilde{S}_{\sigma\nu} - g_{\rho\nu}\tilde{S}_{\sigma\mu} - g_{\sigma\mu}\tilde{S}_{\rho\nu} + g_{\sigma\nu}\tilde{S}_{\rho\mu}\right)$$
$$- \frac{1}{12}\left(g_{\rho\nu}g_{\sigma\mu} - g_{\rho\mu}g_{\sigma\nu}\right)\tilde{R}. \qquad (7.66)$$

Here $\tilde{S}_{\mu\nu}$ is the tracefree Ricci tensor,

$$\tilde{S}_{\mu\nu} = \tilde{R}_{\mu\nu} - \frac{1}{4}g_{\mu\nu}\tilde{R}, \qquad (7.67)$$

which satisfies $\tilde{S}_\mu{}^\mu = 0$.

Contracting now either Eq. (7.65) or Eq. (7.66) with respect to the indices ρ and μ, we find that the trace of the tensor $\tilde{C}_{\rho\sigma\mu\nu}$ vanishes, $\tilde{C}^\rho{}_{\alpha\rho\beta} = 0$. Hence Eqs. (7.65) and (7.66) express the fact that the tensor $\tilde{R}_{\alpha\beta\gamma\delta}$ decomposes into its irreducible components.

7.3.2 Spinor Formulation

The above results may easily be put into the spinor language. The spinor equivalent to the Yang-Mills field strength $f_{a\mu\nu}$ is given by (see Section 7.2)

$$f_{aAB'CD'} = \sigma^\mu_{AB'}\sigma^\nu_{CD'} f_{a\mu\nu}, \qquad (7.68)$$

$$f_{aAB'CD'} = \epsilon_{AC}\overline{\chi}_{aB'D'} + \chi_{aAC}\,\epsilon_{B'D'}, \qquad (7.69)$$

where $\chi_{aAC} = \frac{1}{2}\epsilon^{B'D'} f_{aAB'CD'}$. The spinor equivalent to the tensor ${}^*f_{a\mu\nu}$, the dual to $f_{a\mu\nu}$, is given by

$${}^*f_{aAB'CD'} = i\left(\epsilon_{AC}\overline{\chi}_{aB'D'} - \chi_{aAC}\,\epsilon_{B'D'}\right). \qquad (7.70)$$

Subsequently the spinors equivalent to the tensors $R_{\mu\nu\rho\sigma}$ and $R^\star_{\mu\nu\rho\sigma}$ may be found. So may the spinor equivalent to $f^+_{a\rho\sigma}$ which, by Eq. (7.34), is given by

$$f^+_{aAB'CD'} = 2\chi_{aAC}\,\epsilon_{B'D'}. \qquad (7.71)$$

As a result, the spinor equivalent to the tensor $R_{\mu\nu\rho\sigma}$ is given by

$$R_{AB'CD'EF'GH'} = -f_{aAB'CD'} f_{aEF'GH'}. \qquad (7.72)$$

Using now Eq. (7.69), we then obtain

$$R_{AB'CD'EF'GH'} = -(\xi_{ACEG}\,\epsilon_{B'D'}\,\epsilon_{F'H'} + \epsilon_{EG}\zeta_{ACF'H'}\,\epsilon_{B'D'}$$

$$+\epsilon_{AC}\overline{\zeta}_{B'D'EG}\ \epsilon_{F'H'}+\epsilon_{AC}\ \epsilon_{EG}\overline{\xi}_{B'D'F'H'}). \tag{7.73}$$

In Eq. (7.73) the two spinors ξ_{ABCD} and $\zeta_{ABC'D'}$ are defined by

$$\xi_{ABCD}=\chi_{aAB}\chi_{aCD}, \tag{7.74}$$

and

$$\zeta_{ABC'D'}=\chi_{aAB}\overline{\chi}_{aC'D'}, \tag{7.75}$$

respectively.

From the definition of the spinor ξ_{ABCD} we see that it satisfies the following symmetry properties:

$$\xi_{ABCD}=\xi_{BACD}=\xi_{ABDC}=\xi_{CDAB}. \tag{7.76}$$

Hence it can be decomposed into the sum of a totally symmetric spinor η_{ABCD} and a scalar P,

$$\xi_{ABCD}=\eta_{ABCD}+\frac{P}{6}\left(\epsilon_{AC}\ \epsilon_{BD}+\epsilon_{AD}\ \epsilon_{BC}\right). \tag{7.77}$$

Here the scalar P is the trace of the spinor ξ_{ABCD},

$$P=\xi_{AB}{}^{AB}=\frac{1}{4}f_{a\mu\nu}\left(f_a^{\mu\nu}+i^\star f_a^{\mu\nu}\right). \tag{7.78}$$

A simple calculation, moreover, shows that

$$\xi_{AC}{}^{C}{}_{B}=\frac{P}{2}\epsilon_{AB}. \tag{7.79}$$

7.3.3 Comparison with the Gravitational Field

The spinor ξ_{ABCD} resembles in its properties the gravitational field spinor χ_{ABCD}, which combines the Weyl conformal spinor and the Ricci scalar curvature (see Chapter 6). The difference between the two spinors is only in their trace structure, the trace of the gravitational field spinor is $\chi_{AB}{}^{AB}=-R/4$, where R is the Ricci scalar curvature, which is a real quantity. Here, however, the invariant P is a complex function. The role of P in gauge fields, nevertheless, seems to be similar to that of the cosmological constant of general relativity theory.

The spinor η_{ABCD} in Eq. (7.77), on the other hand, is a totally symmetrical spinor in all of its four indices and is given by

$$\eta_{ABCD}=\frac{1}{3}\left(\xi_{ABCD}+\xi_{ACBD}+\xi_{ADBC}\right). \tag{7.80}$$

7.3. THE GEOMETRY OF GAUGE FIELDS

It is therefore completely analogous to the Weyl conformal spinor, and has only five independent complex components: $\eta_0 = \eta_{0000}$, $\eta_1 = \eta_{0001}$, $\eta_2 = \eta_{0011}$, $\eta_3 = \eta_{0111}$, $\eta_4 = \eta_{1111}$.

The other spinor $\zeta_{ABC'D'}$ appearing in Eq. (7.73), defined by Eq. (7.75), satisfies the same symmetries that the tracefree Ricci spinor $\phi_{ABC'D'}$ satisfies, namely,

$$\zeta_{ABC'D'} = \zeta_{BAC'D'} = \zeta_{ABD'C'} = \overline{\zeta}_{C'D'AB}. \tag{7.81}$$

It therefore has nine real independent components. The spinor $\zeta_{ABC'D'}$ is, moreover, irreducible. Its physical meaning lies in the fact that it is proportional to the energy-momentum tensor of the Yang-Mills field (see details in Section 7.2).

From the spinor $R_{AB'CD'EF'GH'}$ given by Eq. (7.73) we may define the Ricci spinor $R_{CD'GH'} = R^{EF'}{}_{CD'EF'GH'}$. We then find that

$$R_{CD'GH'} = 2\zeta_{CGD'H'} - \frac{1}{2}\left(P + \overline{P}\right)\epsilon_{CG}\,\epsilon_{D'H'}. \tag{7.82}$$

We also find for the Ricci scalar curvature

$$R = R^{GH'}{}_{GH'} = -2\left(P + \overline{P}\right). \tag{7.83}$$

We then find the following expressions:

$$G_{AB'CD'} = R_{AB'CD'} - \frac{1}{2}\epsilon_{AC}\,\epsilon_{B'D'}R = 2\zeta_{ACB'D'} + \frac{1}{2}\left(P + \overline{P}\right)\epsilon_{AC}\,\epsilon_{B'D'}, \tag{7.84}$$

$$S_{AB'CD'} = R_{AB'CD'} - \frac{1}{4}R\epsilon_{AC}\,\epsilon_{B'D'} = 2\zeta_{ACB'D'}, \tag{7.85}$$

for the Einstein spinor and the tracefree Ricci spinor, respectively.

We now find the spinor equivalent to the tensor $R^{\star}_{\alpha\beta\gamma\delta}$ defined by Eq. (7.59). It is given by

$$R^{\star}_{aAB'CD'EF'GH'} = -f_{aAB'CD'}\,{}^{\star}f_{aEF'GH'}. \tag{7.86}$$

Using Eqs. (7.69) and (7.70) we obtain

$$R^{\star}_{aAB'CD'EF'GH'} = i(\xi_{ACEG}\epsilon_{B'D'}\,\epsilon_{F'H'} - \epsilon_{EG}\zeta_{ACF'H'}\,\epsilon_{B'D'}$$

$$+\epsilon_{AC}\overline{\zeta}_{B'D'EG}\,\epsilon_{F'H'} - \epsilon_{AC}\,\epsilon_{EG}\overline{\xi}_{B'D'F'H'}). \tag{7.87}$$

7.3.4 Ricci and Einstein Spinors

The Ricci spinor, Ricci scalar curvature, Einstein spinor, and tracefree Ricci spinor are subsequently given by

$$R^\star_{CD'GH'} = \frac{i}{2}\left(P - \overline{P}\right)\epsilon_{CG}\,\epsilon_{D'H'}, \tag{7.88}$$

$$R^\star = 2i\left(P - \overline{P}\right), \tag{7.89}$$

$$G^\star_{AB'CD'} = -\frac{i}{2}\left(P - \overline{P}\right)\epsilon_{AC}\,\epsilon_{B'D'}, \tag{7.90}$$

$$S^\star_{AB'CD'} = 0, \tag{7.91}$$

respectively.

Finally, the spinor equivalent to the complex tensor $\tilde{R}_{\alpha\beta\gamma\delta}$, defined by Eq. (7.62), is given by

$$\tilde{R}_{AB'CD'EF'GH'} = R_{AB'CD'EF'GH'} + iR^\star_{AB'CD'EF'GH'}. \tag{7.92}$$

We then find that

$$\tilde{R}_{AB'CD'EF'GH'} = -2\left(\xi_{ACEG}\,\epsilon_{B'D'} + \epsilon_{AC}\overline{\zeta}_{B'D'EG}\right)\epsilon_{F'H'}. \tag{7.93}$$

The Ricci spinor, Ricci scalar curvature, Einstein spinor, and tracefree Ricci spinor are then given by

$$\tilde{R}_{CD'GH'} = 2\zeta_{CGD'H'} - P\epsilon_{CG}\,\epsilon_{D'H'}, \tag{7.94}$$

$$\tilde{R} = -4P, \tag{7.95}$$

$$\tilde{G}_{CD'GH'} = 2\zeta_{CGD'H'} + P\epsilon_{CG}\,\epsilon_{D'H'}, \tag{7.96}$$

$$\tilde{S}_{CD'GH'} = 2\zeta_{CGD'H'}, \tag{7.97}$$

respectively.

A fourth spinor which can be constructed out of the Yang-Mills spinor is given by

$$\chi_{ABCDEF} = \epsilon_{abc}\chi_{aAB}\chi_{bCD}\chi_{cEF}. \tag{7.98}$$

It satisfies the following symmetry:

$$\chi_{ABCDEF} = \chi_{BACDEF} = \chi_{ABDCEF} = \chi_{ABCDFE}. \tag{7.99}$$

In addition, the spinor χ_{ABCDEF} keeps or changes its sign, depending upon whether the pairs of indices AB, CD, EF are an even or an odd permutation of the pairs of numbers 00, 01(=10), 11, and zero otherwise. Hence it can be decomposed as follows:

$$\chi_{ABCDEF} = \frac{Q}{24}(\epsilon_{AC}\,\epsilon_{BE}\,\epsilon_{DF}+\epsilon_{AF}\,\epsilon_{BC}\,\epsilon_{DE}+\epsilon_{AC}\,\epsilon_{BF}\,\epsilon_{DE}+\epsilon_{AE}\,\epsilon_{BC}\,\epsilon_{DF}$$

$$+\epsilon_{AD}\,\epsilon_{BF}\,\epsilon_{CE}+\epsilon_{AD}\,\epsilon_{BE}\,\epsilon_{CF}+\epsilon_{AF}\,\epsilon_{BD}\,\epsilon_{CE}+\epsilon_{AE}\,\epsilon_{BD}\,\epsilon_{CF}), \quad (7.100)$$

where Q is a complex quantity, the trace of the spinor χ_{ABCDEF}:

$$Q = \chi_A{}^C{}_C{}^E{}_E{}^A = \epsilon^{CB}\epsilon^{ED}\epsilon^{AF}\chi_{ABCDEF}. \quad (7.101)$$

Finally, two more mixed-indices spinors, with unprimed and primed indices, can be defined as follows:

$$\phi_{ABCDE'F'} = \epsilon_{abc}\chi_{aAB}\chi_{bCD}\overline{\chi}_{cE'F'}, \quad (7.102)$$

$$\phi_{ABC'D'E'F'} = \epsilon_{abc}\chi_{aAB}\overline{\chi}_{bC'D'}\overline{\chi}_{cE'F'}. \quad (7.103)$$

The relationship between them can easily be shown to be given by

$$\phi_{ABCDE'F'} = \overline{\phi}_{E'F'ABCD}, \quad (7.104)$$

$$\phi_{ABC'D'E'F'} = \overline{\phi}_{C'D'E'F'AB}, \quad (7.105)$$

7.4 References for Further Reading

M. Carmeli, Classification of classical Yang-Mills fields, in: *Differential Geometrical Methods in Mathematical Physics*, K. Bleuler, H.R. Petry and A. Reetz, Editors, 105-149 (Springer-Verlag, Heidelberg, 1978). (Sections 7.2, 7.3)

M. Carmeli, *Classical Fields: General Relativity and Gauge Theory* (John Wiley, 1982).

M. Carmeli and M. Fischler, Classification of SU(2) gauge fields: Lorentz-invariant versus gauge-invariant schemes, *Phys. Rev.* D **19**, 3653-3659 (1972). (Sections 7.2, 7.3)

L.P. Eisenhart, *Riemannian Geometry* (Princeton University Press, New Jersey, 1949). (Section 7.3)

R. Jackiw, C. Nohl and C. Rebbi, Conformal properties of pseudoparticle configurations, *Phys. Rev.* D **15**, 1642-1646 (1977). (Section 7.2)

R. Jackiw and C. Rebbi, Conformal properties of a Yang-Mills pseudoparticle, *Phys. Rev.* D **14**, 517-523 (1976). (Section 7.2)

R. Jackiw and C. Rebbi, Spinor analysis of Yang-Mills theory, *Phys. Rev.* D **16**, 1052-1060 (1977). (Section 7.2)

C.N. Yang and R.L. Mills, Conservation of isotopic spin and isotopic gauge invariance, *Phys. Rev.* **96**, 191-195 (1954). (Section 7.1)

Chapter 8

The Euclidean Gauge Field Spinors

In the last chapter we discussed the gauge field spinors in the flat Minkowskian spacetime. In this chapter we extend the discussion to the Euclidean gauge field spinors. First we discuss the Euclidean spacetime in general terms. This is followed by writing down the Dirac equation in this spacetime, and discussing the matrices involved in this formalism. The spinor formulation of the Euclidean gauge fields is subsequently given. This includes the O(4) two-component spinors. The chapter ends with the discussion of the self-dual and anti-self-dual fields appearing in the theory.

8.1 Euclidean Spacetime

We are now in a position to formulate the Euclidean gauge field theory for isospin in terms of quantities which are multispinors of the group product O(4)×SU(2), where O(4) is the four-dimensional rotation group. The group O(4) replaces here the group SL(2,C) employed in the previous chapters for gauge fields and gravitation. Hence instead of dealing with quantities defined in the Minkowskian or the Riemannian spacetimes, as has been done so far, we will be dealing with quantities defined in the Euclidean four-dimensional spacetime.

The Groups O(4) and SU(2)×SU(2)

We will also use the fact that the group O(4) may be written as the product of two SU(2) groups, O(4)=SU(2)×SU(2). Hence in the spinorial formulation described below the O(4) quantities, with which we are concerned in the four-dimensional Euclidean spacetime, may be designated by SU(2)×SU(2) representation labels. Moreover, the additional internal isospin space SU(2) gauge group also gives rise to such SU(2) labels. Accordingly all quantities of interest in the four-dimensional Euclidean spacetime with SU(2) internal symmetry are actually three SU(2) multispinor quantities. A certain simplification is achieved sometimes when the various SU(2) groups are coupled to each other.

The spinorial method makes use of 2×2 matrices to be described below which are Euclidean analogues of the 2×2 SL(2,C) matrices encountered in the study of the Lorentz group. We subsequently present the spinorial formulation of the SU(2) gauge theory. But we first present the Dirac equation in the Euclidean spacetime.

8.1.1 The Euclidean Dirac Equation

Our starting point is the four-dimensional Euclidean spacetime gauge covariant Dirac equation

$$\gamma^\mu \psi_{|\mu} = \gamma^\mu \left(\partial_\mu - ig B_\mu\right) \psi = 0. \tag{8.1}$$

Here $\psi_{|\mu}$ is a gauge covariant derivative of the spinor ψ. Under the infinitesimal SU(2) transformation with generators θ_a, where $a = 1, 2, 3$, the spinor ψ transforms according to some representation of the group SU(2), namely,

$$\delta\psi = iT_a \psi \theta_a. \tag{8.2}$$

The matrices T_a describe the infinitesimal generators of the group SU(2) and satisfy

$$[T_a, T_b] = i\epsilon_{abc} T_c, \tag{8.3}$$

with $a, b, c = 1, 2, 3$. In Eq. (8.1) B_μ is the Yang-Mills gauge potential in a Hermitian matrix representation given by

$$B_\mu = b_{a\mu} T_a. \tag{8.4}$$

8.1. EUCLIDEAN SPACETIME

Gauge Potential and Gauge Field

We assume that from the gauge potential B_μ we can define a well-behaved gauge field strength matrix

$$F_{\mu\nu} = \partial_\nu B_\mu - \partial_\mu B_\nu + i[B_\mu, B_\nu], \qquad (8.5)$$

so that the action integral is finite. This requirement then implies that the *Pontrjagin index*, defined by

$$q = \frac{1}{32\pi^2} \int {}^*f_{a\mu\nu} f_a^{\mu\nu} d^4x, \qquad (8.6)$$

is an integer. Here ${}^*f_{a\mu\nu}$ is the dual to $f_{a\mu\nu}$, and $f_{a\mu\nu}$ is the gauge field strength. Also, in Eq. (8.5) and the rest of this chapter the coupling constant g is taken as unity.

Now ψ is also a four-component spinor in the Euclidean space. The 4×4 Dirac matrices γ^μ satisfy the Euclidean anticommutation relations

$$\{\gamma^\mu, \gamma^\nu\} = 2\delta^{\mu\nu}. \qquad (8.7)$$

The metric here is $\delta_{\mu\nu}$ and the signature is $(+,+,+,+)$ so that there is no distinction between upper and lower space indices.

The γ Matrices

A convenient realization of the γ matrices is given by

$$\gamma_k = \begin{pmatrix} 0 & -i\sigma_k \\ i\sigma_k & 0 \end{pmatrix}, \quad \gamma_4 = \begin{pmatrix} 0 & I \\ I & 0 \end{pmatrix}, \qquad (8.8)$$

where σ_k ($k = 1, 2, 3$) are the three Pauli matrices given by Eq. (7.51) and I is the unit 2×2 matrix. With the above realization for the matrices γ^μ, we define the matrix γ_5 by

$$\gamma_5 = \gamma_1 \gamma_2 \gamma_3 \gamma_4 = \begin{pmatrix} -I & 0 \\ 0 & +I \end{pmatrix}. \qquad (8.9)$$

The matrix γ_5 consequently anticommutates with the matrices γ_μ,

$$\{\gamma_\mu, \gamma_5\} = 0, \qquad (8.10)$$

with $\mu = 1, 2, 3, 4$. As a result, γ_5 also anticommutates with the Hermitian Dirac differential operator $i\gamma^\mu(\partial_\mu - iB_\mu)$.

Hence if we consider the full eigenvalue spectrum through the equation of motion

$$i\gamma^{\mu}\left(\partial_{\mu}-iB_{\mu}\right)\psi_{E}=E\psi_{E}, \tag{8.11}$$

then the matrix γ_5 transforms the spinor ψ_E into the spinor ψ_{-E}, namely,

$$i\gamma^{\mu}\left(\partial_{\mu}-iB_{\mu}\right)\psi_{-E}=-E\psi_{-E}, \tag{8.12}$$

with $\psi_{-E} = \gamma_5\psi_E$. On the other hand, the zero-eigenvalue modes may be chosen as the eigenstates of γ_5, and they have either positive or negative chirality.

8.1.2 Algebra of the Matrices s_μ

Equations (8.8) show that the γ matrices may also be presented in the form

$$\gamma_\mu = \begin{pmatrix} 0 & s_\mu \\ s_\mu^\dagger & 0 \end{pmatrix}, \tag{8.13}$$

where the matrices s_μ and $s_\mu^\dagger$ are defined by

$$s_\mu = \frac{1}{\sqrt{2}}\left(-i\sigma_k, I\right), \tag{8.14}$$

$$s_\mu^\dagger = \frac{1}{\sqrt{2}}\left(i\sigma_k, I\right), \tag{8.15}$$

These 2×2 matrices will be used in the sequel, and some of their properties are given in the following.

We first have

$$s_\mu^\dagger s_\nu + s_\nu^\dagger s_\mu = \delta_{\mu\nu}, \tag{8.16a}$$

$$s_\mu s_\nu^\dagger + s_\nu s_\mu^\dagger = \delta_{\mu\nu}. \tag{8.16b}$$

Hermitian Spin Matrices

We then define the Hermitian *spin matrices*

$$s_{\mu\nu} = \frac{1}{2i}\left(s_\mu^\dagger s_\nu - s_\nu^\dagger s_\mu\right), \tag{8.17a}$$

$$s_{\mu\nu}^\dagger = \frac{1}{2i}\left(s_\mu s_\nu^\dagger - s_\nu s_\mu^\dagger\right), \tag{8.17b}$$

which satisfy

$$2s_\mu^\dagger s_\nu = \delta_{\mu\nu} + 2is_{\mu\nu}, \tag{8.18a}$$

8.1. EUCLIDEAN SPACETIME

$$2s_\mu s_\nu^\dagger = \delta_{\mu\nu} + 2is_{\mu\nu}^\dagger. \tag{8.18b}$$

One then finds that in terms of the Pauli matrices we have

$$s_{ij} = s_{ij}^\dagger = -\frac{1}{2}\epsilon_{ijk}\sigma_k, \tag{8.19a}$$

$$s_{i4} = -s_{i4}^\dagger = \frac{1}{2}\sigma_i, \tag{8.19b}$$

with $i, j, k = 1, 2, 3$ and $\epsilon_{123} = +1$. Moreover, under the duality transformation we have

$$^\star s_{\mu\nu} = -s_{\mu\nu}, \quad ^\star s_{\mu\nu}^\dagger = s_{\mu\nu}^\dagger, \tag{8.20}$$

where

$$^\star s_{\mu\nu} = \frac{1}{2}\epsilon_{\mu\nu\rho\sigma}s^{\rho\sigma},$$

$$^\star s_{\mu\nu}^\dagger = \frac{1}{2}\epsilon_{\mu\nu\rho\sigma}s^{\dagger\rho\sigma}.$$

Commutation Relations

The spin matrices satisfy the O(4) commutation relations given by

$$i\left[s_{\mu\alpha}, s_{\nu\beta}\right] = \delta_{\nu\alpha}s_{\mu\beta} - \delta_{\mu\nu}s_{\alpha\beta} + \delta_{\alpha\beta}s_{\nu\mu} - \delta_{\mu\beta}s_{\nu\alpha}, \tag{8.21a}$$

$$i\left[s_{\mu\alpha}^\dagger, s_{\nu\beta}^\dagger\right] = \delta_{\nu\alpha}s_{\mu\beta}^\dagger - \delta_{\mu\nu}s_{\alpha\beta}^\dagger + \delta_{\alpha\beta}s_{\nu\mu}^\dagger - \delta_{\mu\beta}s_{\nu\alpha}^\dagger, \tag{8.21b}$$

Moreover, the products of three s matrices are equal to linear combination of s matrices,

$$s_{\mu\nu}^\dagger s_\alpha = \left(s_\alpha^\dagger s_{\mu\nu}\right)^\dagger = \frac{1}{2i}\left(\delta_{\nu\alpha}s_\mu - \delta_{\mu\alpha}s_\nu + \epsilon_{\mu\nu\alpha\beta}s^\beta\right), \tag{8.22a}$$

$$s_{\mu\nu}s_\alpha^\dagger = \left(s_\alpha s_{\mu\nu}^\dagger\right)^\dagger = \frac{1}{2i}\left(\delta_{\nu\alpha}s_\mu^\dagger - \delta_{\mu\alpha}s_\nu^\dagger - \epsilon_{\mu\nu\alpha\beta}s^{\dagger\beta}\right). \tag{8.22b}$$

Finally, the following identities

$$\sigma_k s_{\mu\nu}\sigma_k = -s_{\mu\nu} \quad \text{(no summation on } k\text{)}, \tag{8.23a}$$

$$\sigma_k s_{\mu\nu}^\dagger \sigma_k = -s_{\mu\nu}^\dagger \quad \text{(no summation on } k\text{)}, \tag{8.23b}$$

$$s_\mu^\dagger \sigma_k s^\mu = 0, \tag{8.23c}$$

between the Pauli matrices and the s matrices, may be verified.

The eigenvalue equation (8.11) may be written in terms of the s matrices also. It then has the form

$$\begin{pmatrix} 0 & L \\ L^\dagger & 0 \end{pmatrix} \begin{pmatrix} \psi_E^+ \\ \psi_E^- \end{pmatrix} = E \begin{pmatrix} \psi_E^+ \\ \psi_E^- \end{pmatrix}. \qquad (8.24)$$

This is a two-component spinor form which exhibits the chiral structure. Here the operators L and $L^\dagger$ are defined by

$$L = s^\mu (i\partial_\mu + B_\mu), \qquad (8.25a)$$

$$L^\dagger = s^{\mu\dagger} (i\partial_\mu + B_\mu), \qquad (8.25b)$$

The zero-eigenvalue modes then satisfy the equations

$$L\psi^- = 0, \qquad L^\dagger \psi^+ = 0. \qquad (8.26)$$

Here ψ^+ and ψ^- are two-component spinors which also carry an isospin label according to the representation (8.3).

8.2 The Euclidean Gauge Field Spinors

We are now in a position to present the spinorial formulation of the Euclidean Yang-Mills theory with the internal gauge group SU(2). Accordingly all quantities will have spinor indices $A, B, C, \cdots$ taking the values 0,1. As has been mentioned above, these two-component spinors represent the two SU(2) groups in terms of which the Euclidean group O(4) is presented according to O(4)=SU(2)×SU(2).

8.2.1 O(4) Two-Component Spinors

An O(4) two-component spinor is accordingly denoted by ζ^A when having one superscript index or by ζ_A when having one subscript index. The spinor equivalent to an O(4) tensor is obtained like in the case of the group SL(2,C) (see Section 5.2). To each tensorial O(4) index $\mu, \nu, \cdots$, there correspond two spinorial indices which are now denoted by $AA', BB', \cdots$. The spinor equivalent to the vector V_μ is denoted by $V_{AA'}$ and that equivalent to the tensor $T_{\mu\nu}$ is denoted by $T_{AA'BB'}$, for instance. The tensors and spinors are related by means of the s matrices. The vector V_μ and the tensor $T_{\mu\nu}$, on the other hand, may be recovered from the spinors $V_{AA'}$ and $T_{AA'BB'}$ by using the properties of the s matrices.

8.2. THE EUCLIDEAN GAUGE FIELD SPINORS

Matrix Elements

We denote the matrix elements of the matrix s^μ by

$$(s^\mu)_{...} = s^\mu_{AA'}. \tag{8.27}$$

We also denote the matrix elements of the matrix $s^\dagger_\mu$ by

$$(s^\dagger_\mu)_{...} = s^{\dagger A'A}_\mu. \tag{8.28}$$

Accordingly we have

$$s^1_{AA'} = \frac{1}{\sqrt{2}} \begin{pmatrix} 0 & -i \\ -i & 0 \end{pmatrix}, \tag{8.29a}$$

$$s^2_{AA'} = \frac{1}{\sqrt{2}} \begin{pmatrix} 0 & 1 \\ -1 & 0 \end{pmatrix}, \tag{8.29b}$$

$$s^3_{AA'} = \frac{1}{\sqrt{2}} \begin{pmatrix} -i & 0 \\ 0 & i \end{pmatrix}, \tag{8.29c}$$

$$s^4_{AA'} = \frac{1}{\sqrt{2}} \begin{pmatrix} 1 & 0 \\ 0 & 1 \end{pmatrix}, \tag{8.29d}$$

and

$$s^{\dagger A'A}_1 = \frac{1}{\sqrt{2}} \begin{pmatrix} 0 & i \\ i & 0 \end{pmatrix}, \tag{8.30a}$$

$$s^{\dagger A'A}_2 = \frac{1}{\sqrt{2}} \begin{pmatrix} 0 & -1 \\ 1 & 0 \end{pmatrix}, \tag{8.30b}$$

$$s^{\dagger A'A}_3 = \frac{1}{\sqrt{2}} \begin{pmatrix} i & 0 \\ 0 & -i \end{pmatrix}, \tag{8.30c}$$

$$s^{\dagger A'A}_4 = \frac{1}{\sqrt{2}} \begin{pmatrix} 1 & 0 \\ 0 & 1 \end{pmatrix}. \tag{8.30d}$$

Superscript and Subscript Indices

There is no distinction between the superscript and the subscript O(4) tensorial indices, namely, $(s^\mu)_{...} = (s_\mu)_{...}$ and $(s^{\mu\dagger})_{...} = (s^\dagger_\mu)_{...}$. The spinor equivalent to the vector V_μ is thus given by

$$V_{AA'} = s^\mu_{AA'} V_\mu$$

$$= \frac{1}{\sqrt{2}} \begin{pmatrix} -iV_3 + V_4 & -iV_1 + V_2 \\ -iV_1 - V_2 & iV_3 + V_4 \end{pmatrix}. \tag{8.31}$$

The spinorial indices $A, A', \cdots$ may be raised or lowered according to the ordinary rules for the SL(2,C) spinors, given in Subsection 5.2.2, namely,

$$\zeta^A = \epsilon^{AB}\zeta_B, \quad \zeta_A = \zeta^B \epsilon_{BA}. \tag{8.32}$$

Hence if we define the spinors

$$s_\mu^{AA'} = \epsilon^{AB}\epsilon^{A'B'} s_{\mu BB'}, \tag{8.33}$$

$$s_{\mu A'A}^\dagger = s_\mu^{\dagger B'B} \epsilon_{B'A'}\epsilon_{BA}, \tag{8.34}$$

we then find that

$$s_\mu^{AA'} = s_\mu^{\dagger A'A} = \left(s_\mu^\dagger\right)_{\cdots}, \tag{8.35}$$

$$s_{\mu A'A}^\dagger = s_{\mu AA'} = (s_\mu)_{\cdots}. \tag{8.36}$$

From Eqs. (8.16) we also obtain

$$s_\mu^{AA'} s_{\nu AB'} + s_\nu^{AA'} s_{\mu AB'} = \delta_{B'}^{A'}\delta_{\mu\nu}, \tag{8.37}$$

$$s_{\mu AA'} s_\nu^{BA'} + s_{\nu AA'} s_\mu^{BA'} = \delta_A^B \delta_{\mu\nu}, \tag{8.38}$$

From the above formulas one then obtains the following:

$$s_{\mu AA'} s_\nu^{AA'} = \delta_{\mu\nu}, \tag{8.39}$$

$$s_\mu^{AA'} s_{BB'}^\mu = \delta_B^A \delta_{B'}^{A'}, \tag{8.40}$$

$$s_{\mu AA'} s_{BB'}^\mu = \epsilon_{AB}\epsilon_{A'B'}. \tag{8.41}$$

Using now Eq. (8.40), for instance, we obtain

$$V_{AA'} s_\mu^{AA'} = V_\alpha s_{AA'}^\alpha s_\mu^{AA'} = V_\alpha \delta_\mu^\alpha = V_\mu, \tag{8.42}$$

for the relationship between an O(4) vector and its spinor equivalent.

8.2. THE EUCLIDEAN GAUGE FIELD SPINORS

8.2.2 Self-Dual and Anti-Self-Dual Fields

We may now write the spinors for the gauge potential b^a_μ and the gauge field strength $f^a_{\mu\nu}$ in the Euclidean space. They are given, respectively, by

$$b_{MNAA'} = b^a_\mu \sigma_{aMN} s^\mu_{AA'}, \qquad (8.43)$$

and

$$f_{MNAA'BB'} = f^a_{\mu\nu} \sigma_{aMN} s^\mu_{AA'} s^\nu_{BB'}, \qquad (8.44)$$

where σ_{aMN} are given by Eqs. (7.54). The pair of indices MN are internal SU(2) spinor indices, whereas AA' and BB' are O(4) spinor indices. Both the gauge potential and the gauge field strength are symmetric in their SU(2) spinor indices M and N.

Obviously the field strength spinor (8.44) is skew-symmetric in the pair of indices AA' and BB'. The above spinors are related by

$$f_{MNAA'BB'} = \partial_{BB'} b_{MNAA'} - \partial_{AA'} b_{MNBB'} + 2b_{(MPAA'} b^P{}_{N)BB'}, \qquad (8.45)$$

where brackets indicate symmetrization,

$$\zeta_{(AB)} = \frac{1}{2}(\zeta_{AB} + \zeta_{BA}), \qquad (8.46)$$

and the differential operator $\partial_{AA'} = s^\mu_{AA'} \partial_\mu$.

Now because of its antisymmetrical property, the gauge field strength spinor can be split into two parts as follows:

$$f_{MNAA'BB'} = \epsilon_{AB} f^+_{MNA'B'} + f^-_{MNAB} \epsilon_{A'B'}, \qquad (8.47)$$

where

$$f^+_{MNA'B'} = \frac{1}{2} f_{MNAA'}{}^A{}_{B'}, \qquad (8.48a)$$

$$f^-_{MNAB} = \frac{1}{2} f_{MNAA'B}{}^{A'}. \qquad (8.48b)$$

Here $f^+_{MNA'B'}$ is symmetric under the exchange of the indices A' and B' since

$$f^+_{MNB'A'} = \frac{1}{2} f_{MNAB'}{}^A{}_{A'} = -\frac{1}{2} f_{MN}{}^A{}_{A'AB'} = \frac{1}{2} f_{MNAA'}{}^A{}_{B'} = f^+_{MNA'B'}. \qquad (8.49)$$

Likewise, f^-_{MNAB} is symmetric under the exchange of the indices A and B.

Finally, from Eq. (8.45) it follows that

$$f^+_{MNA'B'} = \partial_{A(A'} b_{MN}{}^A{}_{B')} + b_{(MPAA'} b^P{}_{N)}{}^A{}_{B'}, \qquad (8.50a)$$

$$f^-_{MNAB} = \partial_{(AA'}b_{MNB)}{}^{A'} + b_{(MPAA'}b^P{}_{N)B}{}^{A'}. \qquad (8.50b)$$

Furthermore, under the duality transformation we find

$$^\star f_{MNAA'BB'} = f_{MNAB'BA'}, \qquad (8.51)$$

which in terms of f^+ and f^- can be written as

$$^\star f^+_{MNA'B'} = f^+_{MNA'B'}, \qquad (8.52a)$$

$$^\star f^-_{MNAB} = -f^-_{MNAB}. \qquad (8.52b)$$

Hence for self-dual fields the expression f^-_{MNAB} must vanish, whereas for anti-self-dual fields the expression $f^+_{MNA'B'}$ must vanish.

8.3 Problems

8.1 Verify Eqs. (8.16), (8.17) and (8.18).

Solution: These equations are direct consequences of the definition of the s matrices and are left to the reader for verification.

8.2 Verify Eqs. (8.21) and (8.22).

Solution: Equations (8.21) and (8.22) are left to the reader for verification.

8.4 References for Further Reading

M.F. Atiyah, N.J. Hitchin and I.M. Singer, *Proc. Nat. Acad. Sci. (USA)* **74** (1977). (Section 8.1)

M.F. Atiyah, V. Patodi and L. Singer, *Math. Proc. Camb. Philos. Soc.* **77**, 43 (1975); **78**, 405 (1975); **79**, 71 (1976). (Section 8.1)

L.S. Brown, R.D. Carlitz and C. Lee, Massless excitations in pseudoparticle fields, *Phys. Rev. D* **16**, 417-422 (1977). (Section 8.1)

M. Carmeli, *Classical Fields: General Relativity and Gauge Theory* (John Wiley, 1982).

R. Jackiw, C. Nohl and C. Rebbi, Conformal properties of pseudoparticle configurations, *Phys. Rev. D* **15**, 1642-1646 (1977). (Section 8.1)

R. Jackiw and C. Rebbi, Conformal properties of a Yang-Mills pseudoparticle, *Phys. Rev. D* **14**, 517-523 (1976). (Section 8.1)

8.4. REFERENCES FOR FURTHER READING

R. Jackiw and C. Rebbi, Spinor analysis of Yang-Mills theory, *Phys. Rev. D* **16**, 1052-1060 (1977). (Sections 8.1, 8.2)

F.R. Ore, Jr., Quantum field theory about a Yang-Mills pseudoparticle, *Phys. Rev. D* **15**, 470-479 (1977). (Section 8.1)

Index

(Page numbers in italics refer to publications cited in the references.)

Absolute rest, 36
Acceleration, 118, 119, 139
    gravitational, 118
    radial, 126
Action integral, 121, 187
Angles, Euler, 4-6
Angular momentum,
    isotopic spin, 170
    per mass unit, 129
Approach,
    infinitesimal, 47-54
Atiyah, M.F., *194*
Atoms, 119
    frequency of, 127
    frequency shift of, 127, 129
    gold and aluminium, 119
Automorphism, 3

Bade, W.L., *107, 163*
Banach space, 27-29
    operators in, 30
Barut, A.O., *62*
Basis, 12, 13, 17, 52
    canonical, 16-17, 20, 23, 52
Bertotti, B., *163*

Bhattacharjee, R., *107*
Bianchi identities, 117-118, 135, 147, 157
    contracted, 118, 121, 134, 147
    tensorial, 147
Birkhoff, G., 127, *163*
Bjorken, J.D., *107*
Bleuler, K., *183*
Boener, H., *81*
Bohm, D., *62*
Boost, 41, 47
Born, M., *62*
Brauer, R., *82*
Brill, D., *163*
Brown, L.S., *194*

Carlitz, R.D., *194*
Carmeli, M., *10, 32, 62, 63, 82, 107,* 135, *163, 164, 183, 194*
Cartan, E., 65, *82*
Cauchy-Buniakovsky inequality, 28
Cauchy condition, 28
Charge, 99
    density of, 85
    states of,
        nucleon, 168
Charged particles,
    Lagrangian density of, 84
Chaudhury, T., *107*
Chevalley, C., *10, 82*

Chirality, 188, 190
Christoffel symbols, 113-114, 117,
    125, 135, 136
Circumference, 123
Clocks, 38, 127
    atomic, 128
Cohen, J.M., *107*
Collinson, C., *105*
Condition, Cauchy, 28
Conditions,
    unitary, 53-54
Cone,
    galaxy, 43-44
        symmetry axis of, 44
    light, 44, 91
    negative, 60
    positive, 60, 61
Constant,
    coupling, 187
Continuity equation, 172
Coordinate, 42
    Cartesian, 127, 131
    spacetime, 94, 168, 169
    spatial, 36, 38, 41, 44, 45
    spherical, 136, 138
    time, 36, 37, 38, 41, 44
    translation of, 37
Coordinate system, 111, 112, 114,
    118, 120, 121, 124
    Cartesian, 5, 93
    geodesic, 117, 121
    inertial, 36-37, 38, 39, 44
        orientation of, 38
        origin of, 38, 44
Correspondence,
    between isovectors and isospinors,
        176
    between spinors and tensors,
        88
Corson, E.M., *108*

Coset, 3
    left, 3
    right, 3
Cosmic time, 41, 42, 43
Cosmological constant, 121, 180
Cosmology, 43
Coulomb field,
    high frequency, 132
Covariant derivatives, 87, 114-115,
    132, 143, 144, 168, 169
    gauge, 186
    operator of, 93
    ordinary coordinate, 94, 143
    spin, 93-94
    spinor, 92-95
Covariant differentiation, 86, 114-
    115
Cranshaw, T.F., 128, *164*
Cross-section,
    differential, 168
Current density,
    vector, 85

Davis, T.M., *105*
Deflection of light, 130-132
    angle of, 132
Dicke, R.H., 119, *164*
Dirac, P.A.M., *82*, *108*
Dirac delta function, 134, 137
Dirac equation, 98-99, 186
    Euclidean, 186-188
    gauge covariant, 186
Direct sum, 52
Direction, 111
Distance, 43
    Earth-planet, 132
    Earth-Sun, 133
    from the center, 123
    infinitesimal, 136
    planet-Sun, 133

INDEX

Drell, D., *107*

Earth,
    gravitation of, 118
    motion of, 36
    orbit radius of, 128
    Schwarzschild radius of, 126
Eccentricity, 129
Eigenvalue equation, 190
Eigenvalue spectrum, 188
Einstein, 35, 36, *63*, 120, 133, 134, 139, *164*
Einstein-Dirac equations, 105, 106, 107
Einstein-Infeld-Hoffmann equation, 141-142
Einstein-Infeld-Hoffmann method, 138-140
Einstein's field equations, 109, 120-121, 123, 124, 126, 127, 132, 134, 135, 175
    deduction from variational principle, 121-122
    exact solution of, 123-127
    with electromagnetic field, 175-176
Einstein-Maxwell field equations, coupled, 123
Einstein summation convention, 110
Einstein tensor,
    the spinor equivalent to, 153
Eisenhart, L.P., *10*, 144, *164*, 178, *183*
Electric field, 84
Electrodynamics, 36, 172, 175
    laws of, 36
Electromagnetic field, 35, 36, 95, 96, 122, 169, 174, 175
    antisymmetry property of, 96
    Lagrangian density for, 83, 84

Electromagnetic potential, 95, 99
Electromagnetics, 169, 173
Electromagnetic spinor,
    decomposition of, 96-97
Energy and momentum,
    conservation of, 121
Energy-momentum tensor,
    zero, 105
Eötvös, 118, 119
Equations, Einstein-Dirac, 105, 106, 107
Equations of motion, 133-142, 171, 188
    Newtonian, 140-141
Euclidean space, 13, 16, 27, 28, 29, 187, 193
    n-dimensional, 15
Euler angles, 4-6, 8, 9, 14, 17, 21, 22, 27
Events, 44
Examples of spinor representation, 71-73
Experiment,
    Eötvös, 118-119
    gravitational radiation, 132
    low-temperature, 133
    Michelson and Morley, 36, 38
    Null, 118-119
    radar, 132-133

Factor group, 1, 3
Fickler, S.I., *165*
Field,
    anti-self-dual, 193-194
    classical, 36
    Coulomb, 132
    covariant isospin-gauge, 170
    electromagnetic, 35, 36, 169
    gauge invariant, 169
    Maxwell, 87

self-dual, 193-194
spin-$\frac{1}{2}$, 172
Field equations, 86, 122, 134, 140, 169, 171-172
  nonlinearity of, 172
Field strength, 172-177
  gauge, 167, 172, 187, 193
    dual to, 173-174, 178
    spinor equivalent to, 167, 172, 192
  Yang-Mills,
    spinor equivalent to, 179
Fierz, M., *108*
Fischler, M., *183*
Four-current density, 84
Four-vectors, 55
  Minkowskian space of, 55
Fourier transform,
  generalized, 79
Frame,
  accelerated, 119
French, A.P., *63*
Function,
  continuous, 31, 50, 57
  gradient of, 111
  numerical, 29, 31
    continuous, 31
  operator, 31, 50
    continuous, 31
  scalar, 111
  vector, 31, 50
Functional,
  bounded linear, 30, 31
  linear, 29
Functions,
  basis, 22
  Riemannian plane of, 60
  space of, 23

Galaxies,
  distribution of, 43
  locations of, 43
Galaxy cone, 43-44
Galileo, 118
Gauge,
  electromagnetic, 169
  isotopic, 169, 170
Gauge fields, 169, 174, 175, 176, 180, 185, 187
  geometry of, 177-183
Gauge invariance, 167
Gauge theory,
  SU(2),
    spinorial formulation of, 186
Gauss' theorem, 122
Gelfand, I.M., *32*, *33*, 53, *63*, *82*
General relativity,
  experimental tests of, 127-133
  original formulation of, 120
General relativity theory, 88, 119, 124, 132, 133, 134, 180
  classical, 120
  elements of, 109-143
  necessary condition of, 118
  sufficient condition, 119
Generators,
  infinitesimal, 14-15
    commutation relations, 16
Geodesic deviation equation, 132
Geodesic equations, 116, 126, 128, 130, 133, 134
Geodesic postulate, 133-134
Geodesics, 116-117, 128, 132
Geometry,
  Riemannian, 110-118, 120
Ghost neutrinos, 105
Goldstein, H., *11*
Graev, M.I., *32*, *33*, *63*, *82*
Gravitation, 120, 185
  Einstein's constant of, 121

INDEX

field of, 119
Newtonian theory, 120, 121
Newton's constant of, 121, 126, 142
potential of, 120
presence of, 87
waves of, 132
Gravitational field, 95, 119, 123, 127, 128, 130, 133, 134, 147, 180-181
  clocks in, 127
  curvature tensor, 95
  deflection of light in, 130
  dynamic, 132
  Earth's, 128
  Einstein's equations of, 132
  equations of, 109, 120-122, 125, 134, 139, 175
    spinor equivalent to, 175
  external, 134
  in vacuum, 126
  Lagrangian for, 121
  slowing down of electromagnetic waves, 132
  Sun's, 128
  weak, 121
Gravitational red shift, 127-128
Green function, 139
Grommer, 134
Group, 1-2
  abelian, 2
  basic one-parameter, 15, 50
  center of, 58, 59
  compact, 22
  factor, 3
  finite, 2
  Galilean, 37
  infinite, 2
  L, 46, 47, 51, 52, 53, 54, 56, 57, 58, 59

infinitesimal matricies of, 48
irreducible representations, 51, 52, 54
linear representation, 49
Lorentz, 35-63, 70, 88, 186
  covering group of, 70
  homogeneous, 46, 56, 57
  inhomogeneous, 46
  orthochronous, 46, 47
  proper, 46, 47, 56, 57
  proper orthochronous, 46, 51, 54, 57, 58, 59
  subgroups of, 46-47
O(4), 185, 186, 190
O(4)×SU(2), 185
order of, 2
Poincaré, 46
pure rotation, 1, 4-6, 13
  three-dimensional, 13, 51
rotation, 1, 7, 48
  four-dimensional, 185
simple, 3
SL(2,C), 35-63, 65, 66, 77, 78, 83, 88, 93, 185, 186, 190
  center of, 58, 59
  inhomogeneous, 59
  representation of, 75, 77, 78, 93
  space of representation, 66
  spinor representation, 65-73, 77, 93
  subgroups of, 59-60
  unity element of, 67
SL(2,C)/Z_2, 59
SL(2,R), 59, 60
spinor representation, 19-20
SO(3), 4-6, 7, 8, 9, 51
  invariant integral over, 8
  representation of, 13-18, 23
SU(1,1), 59, 60

SU(2), 6-7, 65, 78, 169, 190
    complexification of, 76
    generators of, 186
    internal gauge, 190
    invariant integral, 8-9
    irreducible representation of, 76, 79
    operators of, 18
    parametrization of, 59
    representation of, 13-18, 19-20, 21, 76, 186
    SU(2) gauge, 186, 190
    SU(2)×SU(2), 186, 190
    subgroup of, 3
    translational T_4, 59
    triangular matrices, 60
    U(1)×T_2, 60
    unimodular unitary, 13
Group element, 1, 2
    Hermitian conjugate of, 55, 56
    inverse of, 2
    left inverse of, 2
    negative of, 2
    right inverse of, 2
Group elements, 1, 2, 3, 23
    addition of, 2
    central, 58, 59
    equivalent, 3
    multiplication of, 2
    product of, 2, 25
Group theory, introduction to, 1-10
Gyroscope,
    anomalous precession of, 133

Harish-Chandra, 53, *63*
Heisenberg, 168
Hilbert space, 29-30
l_2^{2s}, 79-80
$L_2^{2s}(SU(2))$, 78, 79, 80
    orthogonal sum of, 32
Hitchin, N.J., *194*
Hoffmann, 139
Homomorphism, 1, 3-4
    between SO(3) and SU(2), 6-7, 13
    between the Lorentz group and SL(2,C), 54, 88
    continuous, 56
    kernel of, 4, 58-59
    natural, 4
    of SL(2,C) on L, 56-58
Hubble's constant, 41
Hubble's time, 41
Hyperboloid,
    single-sheeted, 60, 61
    two-sheeted, 60, 61

Iaglom, A.M., 53, *63*
Identity, 2
    left, 2
    right, 2
Inertia,
    law of, 36
Infeld, L., 88, *108*, 134, 139, *164*
Infinite-dimensional spinors, 77-80
Interactions,
    electromagnetic, 168
    $n-p$, 168
    nucleon-nucleon, 168
    pion-nucleon, 168
    $p-p$, 168
    strong, 168
Interval, 123
Intrinsic spin structure, 97-98
Invariance,
    Galilean, 38
    Lorentz, 38

INDEX

Invariant, 111, 112
Invariant integral, 1, 8-9
   over SO(3), 8
   over SU(2), 9, 14
Inverse, 2
   left, 2
   right, 2
Isomorphism, 1, 3-4
   between L and $SL(2,C)/Z_2$, 58-59
   natural, 4
   standard, 60
Isospin, 185, 190
Isospinor, 176
Isovector, 176
   spinor equivalent to, 176

Jackiw, R., *183, 184, 194, 195*
Jackson, J.D., *108*
Jacobian, 110, 112
Jehle, J., *107, 163*

Kahan, T., *108*
Kleinert, H., *62*
Kronecker delta function, 112
Krori, K.D., *107*
Krotkov, R., *163, 164*

Lagrange equation, 117
Lagrangian, 117, 121, 122, 171
Lagrangian density, 83, 84, 86, 122, 171
   charged particles, 83, 84
   electromagnetic field, 83
   isotopic gauge invariant, 171
   total, 171
Laws of nature, 119
Lee, C., *194*
Leibowitz, E., *107, 164*

Length,
   element of, 112
   extremal,
      curves of, 116
Levi-Civita metric spinors, 88, 89, 92
Levi-Civita symbol, 49
L'Hospital's theorem, 137
Line of nodes, 5
Linear operators, 11-12
   addition of, 11
   multiplication of, 11-12
Light,
   deflection of, 127, 130-132
   null propagation of, 42
   propagation of, 42
   propagation velocity, 36
   velocity in empty space, 36
Light pulse, 44
Light signals, 37

Magnetic field, 84
Malin, S., *10, 62, 82*
Mapping, 3, 4
   between $SL(2,C)$ and L, 58
   between $SU(1,1)$ and $SL(2,R)$, 60
Mass, 99, 131, 138, 139, 140, 168
   distance from, 126
   gravitational, 119
   inertial, 119, 135
   of the Sun, 128
   spherical, 133
   with spherical symmetry, 124
Matrices,
   D^j, 21, 22
   Dirac, 187
   γ, 187-188
   group of, 31
   Hermitian, 60, 61, 88, 89, 169

space of, 55, 60
unimodular, 61
infinitesimal, 15, 18, 48-49, 51, 73
   commutation relations of, 15, 16, 49, 51
isospin-gauge covariant field, 170
isotopic spin "angular momentum", 170
Lorentz, 47-48
   infinitesimal, 47-49
orthogonal, 4, 5, 6, 45
Pauli spin, 6, 18, 55, 56, 73, 88, 90, 170, 176, 177, 187, 189
   symmetry of, 176, 177
properties of D^j, 21-22
real orthogonal, 4, 5
σ, 89-90, 92
   covariant derivatives of, 92
s_μ,
   algebra of, 188
   elements of, 191
   products of, 189
SL(2,C), 186
spin, 188
   commutation relations of, 189
unitary, 6, 7, 17, 21
Matrix,
   eigenvalues of, 58
   γ_5, 187
      eigenstates of, 188
   Hermitian, 6, 55, 152
   Hermitian conjugate of, 7, 55, 56
   Λ, 39-41, 55, 56, 57, 58
      inverse, 40
   normalized eigenvector, 58

orthogonal, 5, 14
unit, 45, 55, 88, 90
unitary, 7, 17, 21, 22, 169
Maxwell, 36
Maxwell equations, 83, 84, 85, 87, 98-99, 123
   alternative form of, 85
   generalization into curved spacetime, 86-87
   in the presence of gravitation, 87
   spinor version of, 99
Maxwell field, 87
Maxwell spinor, 97
Maxwell tensor, 86
Maxwell's theory, 36, 83-87
Measure, 8, 9
Measuring rods, 38
Metric, 115, 116, 123, 124, 125, 128, 187
   components of, 126
   conformally flat, 116
   flat-space, 44, 123, 126, 127
   flat spacetime, 148
   g, 90-91
   geometrical, 120
   isometric,
      operator, 31
   Minkowskian, 90-91
   Schwarzschild, 127
   spherical symmetry of, 123, 126
Metric tensor, 88, 89, 112-116, 120, 139
   cofactor of, 112
   contravariant, 124
   covariant, 124
   curved spacetime, 86
   expansion of, 139
   flat spacetime, 90, 148

INDEX

geometrical, 88, 89, 120
   spinor equivalent to, 90, 148
   Minkowskian, 90, 91
   transformation law for, 113
Miller, A.I., *63*
Mills, R.L., 169, 170, *183*, *184*
Minlos, R.A., *33*
Morris, P., *105*
Moses, *33*
Mössbauer effect, 128
Motions, 60, 61
   Euclidean, 60
      transitive group of, 61
   Lobachevskian, 60-61
   planet's, 133
   rectilinear, 37
   spinning bodies, 133
   translational, 37
   uniform, 37
Multiplication,
   associative, 2
   group, 58, 60
   group elements, 2, 3
Multispinor, 185, 186

Naimark, M.A., *10*, *33*, 53, *63*, *82*
von Neumann, J., *108*, *165*
Neutrino equation, 98
Neutrinos,
   ghost, 105
   non-ghost, 107
Neutron, 168, 169, 176
Newton, 118
Newtonian approximation, 139, 140
Newtonian equation, 138-139, 142
Newton's law of motion, 126
Newton's laws of mechanics, 36, 38, 129
Nissani, N., *107*, *164*
Nohl, C., *183*, *194*

Non-ghost neutrinos, 107
Nonrelativistic mechanics, 129
Norm, 27-28, 30, 50
Normal subgroup, 1, 3
Nucleon,
   field of, 168
   light, 168
Null conditions, 130
Null experiment, 118-119
   Michelson-Morley, 38

Operator, 13
   adjoint, 16, 32, 53-54
   angular momentum, 26-27
   basic infinitesimal, 16
   bounded, 30, 32
      norm of, 30
   continuous, 30
   covariant derivative, 93, 157
   differential, 24-26
   eigenvalue of, 52
   normalized eigenvectors of, 52
   Hermitian, 16, 32
      Dirac, 187
   infinitesimal, 25, 49-52, 53, 74-75
   inverse, 32
   isometric, 30
   linear, 13-14, 30, 31, 50
   projection, 32
   skew-symmetric, 16
   spinor,
      matrix elements of, 75-76
   unitary, 13, 16, 32, 53
Operator function, 12
   continuous, 31
Operators,
   angular momentum, 26-27
   basic infinitesimal, 16, 50
   continuous, 12

orthogonal sum of, 32
Ore, F.R., Jr., *195*
Orthogonal sum, 32
"Orthogonality" condition, 39

Palatini formalism, 163
Papapetrou, A., 133, *164*
Parametrization, 9, 16
Particles, 71
    motion of, 140, 141
    system of, 134
    test, 128, 132, 133
    velocities of, 138
Patodi, V., *194*
Pauli, W., *108*
Pauli matrices, 18
Pauli spinor, 98
Penrose, R., *108*, *164*
Perihelion, 130
    orbit, 130
        advance of, 130
Periodicity conditions, 9
Petry, H.R., *183*
Phase factor, 167-169
Pion, 168
    three charge states of, 168
Pirani, F.A.E., *164*
Plane symmetry spacetime, 106
Planetary motion,
    effects of general relativity, 127, 128-130
Planets,
    motion of, 128, 129, 130
Poincaré group, 46
Poisson equation, 121
Polynomials, 19, 69, 71, 72, 73
    addition of, 66
    coefficients of, 71, 72, 77, 79
    expansion coefficients of, 76
    expansion into powers, 75

    homogeneous, 19, 75, 76
    Jacobi, 21
    operators on, 69, 71
    product by a number, 66
    space of, 19, 65-66, 72, 75, 77
        dimension of, 66
Pontrjagin, L., *10*, 22, *33*
Pontrjagin index, 187
Post-Newtonian approximation, 138-140
Potential, 176
    B, 169-171
    b_μ, 169-171
    electromagnetic, 95, 99, 168
        spinor equivalent to, 95
    gauge, 172-177, 187, 193
        spinor equivalent to, 167, 172, 192
        Yang-Mills, 186
    scalar, 85
    vector, 85
Pound, R.V., 128, *164*
Prerelativity, 38
Principle of equivalence, 109, 118-119
Principle of general covariance, 109, 119-120
Principle of relativity, 36, 38
    general, 120
    special, 119, 120
Product, 2, 18, 19
Proton, 168, 169, 176
Pure rotation group, 4-6
Pyatetskii-Shapiro, I.I., *32*

Quadratic differential form, 112

Rabka, G.A., 128, *164*
Radar pulses,
    time delay of, 132-133

INDEX

Radius,
    Earth's orbit, 128
    Schwarzschild, 126
    Sun, 128
Ray,
    light,
        orbit of, 131, 132
    one-index spinor, 91
Ray, J.R., *105*
Rebbi, C., *184, 194, 195*
Rechenick, K.R., *107*
Redshift, 43
Reetz, A., *183*
Reflection, 4
Representation, 12, 13, 30-31
    canonical basis for, 20
    continuous, 12, 13, 31
    dimension of, 12
    double-valued, 13, 14, 21, 58
    finite-dimensional, 11, 20, 50, 52
        of $SL(2,C)$, 68, 77
        of $SU(2)$, 20
    general definition, 30-31
    infinite-dimensional, 11, 49, 50, 51, 52, 77
    infinitesimal operator of, 50, 51
    irreducible, 12, 13, 14, 17, 20, 23, 30, 52, 54, 78
    of L,
        irreducible, 51, 52
    of $SL(2,C)$, 67, 75
        finite-dimensional, 68, 77
        spinor, 73
    of $SO(3)$, 16-21, 23
        finite-dimensional, 16
        irreducible, 17, 23, 51, 52
        uniquely determined, 16, 17
    of $SU(2)$, 13, 14, 17, 18, 19, 21
        irreducible, 79
        spinor, 19-20
        unitary irreducible, 76
    single-valued, 13, 21, 58
    space of, 12, 17, 52, 66, 71, 72, 79, 80
    spinor, 19-20, 20, 65, 68, 69, 70, 71, 77
        dimension of, 68, 71
        examples of, 71-73
        operators of, 73-77
        realization of, 66-68
    unitary, 13, 16, 31-32, 53, 78
    weight of, 17, 19, 20, 23, 51
Representations,
    , double-valued, 13, 14
    finite-dimensional, 12, 16, 77
        equivalent, 12
    infinite-dimensional, 27-32
        equivalent, 58
    matrix elements of, 18-22
    orthogonal sum of, 32
    principle series of, 54, 77-78, 80
    uniquely determined, 16
    unitary, 13
Representation theory, 11-33, 65
Ricci identity,
    spinorial, 144-145
Ricci scalar, 115, 121
Ricci scalar curvature, 123, 142, 152, 153, 156, 175, 178, 181, 182
Ricci tensor, 115-116, 140, 142, 178
    tracefree, 142, 152, 153, 179
    spinor equivalent to, 153
Roll, P.G., *164*

Rotation, 4, 5, 6, 7, 13, 14, 15, 20, 24, 25, 26, 45, 57, 60
    angle of, 13, 24, 25, 27
    differential operator, 22-27
    direction of, 13, 27
    infinitesimal, 22
    infinitesimal matrices, 18
    matrix of, 24, 47
    orthogonal, 37
    pure, 4
    spatial coordinates, 41, 45
    three-dimensional, 4, 41
    unitary matrix, 17-18
Rotation group, 1, 4-6
Rühl, *63*, 75, *82*

Scalar function,
    gradient of, 111
Scalar potential, 85
Scalar product, 13, 27, 29, 30, 53, 55, 78, 79
Schiff, L.I., 133, *164*
Schiffer, S.P., 128, *164*
Schild, A., 134, *164*
Schwarzschild, 123
Schwarzschild field, 123-127, 132, 133
Schwarzschild metric, 127
Schwarzschild radius, 126
    Earth, 126
    electron, 126
    Sun, 126
Schwarzschild solution, 109, 123-127
Self-action terms, 135-138
Sequence,
    convergent in the norm, 28
    fundamental, 28
Series,
    absolutely convergent, 29

    complementary, 54
    convergent, 29
        sum of, 32
    principal, 54
Set,
    base vectors, 52
    closed, 29
        bounded, 31
    closure of, 29
    dense, 29
Shapiro, I.I., 132, 133, *164*
Shapiro, Z.Ya., *33*, *33*
Signature, 187
Simultaneity, 37
Singer, I.M., *194*
Singer, L., *194*
Space, 11, 12
    Banach, 27-29, 30, 31
        norm in, 50
        operators in, 30
        subspace of, 50
    Cartesian, 41
    complete, 30
    conformal, 116
    conjugate, 30, 31
    dimension of, 69
    dual of distance and velocity, 43
    Euclidean, 13, 16, 27, 28, 29, 187, 193
    n-dimensional, 15
    finite-dimensional, 12, 13, 31, 68
    flat, 123, 126, 127, 132
        line element of, 123
    four-dimensional, 42, 110, 115
    4-dimensional flat, 41
    Hilbert, 29-30, 32, 53, 78, 79
    l_2^{2s}, 79-80
    $L_2^{2s}(SU(2))$, 78, 79, 80

INDEX

orthogonal sum of, 32
infinite-dimensional, 31, 49
inner,
    degree of freedom, 172, 176
isospin, 170
    internal, 186
l^2, 28, 29
linear, 11, 19, 27, 29, 30, 66, 67, 68, 75
Lobachevskian, 61
m-dimensional, 31
Minkowskian, 55, 60
non-Euclidean, 135
normed, 27-29, 30, 31
    complete, 28, 29
polynomials, 65-66, 71
    dimension of, 69
    unit operator in, 66
R, 27, 28, 29, 30, 31, 32
reflexive, 31
Riemannian, 123
three-dimensional, 137
Space inversion, 47
Spaces,
    conformal, 116
Spacetime, 169
    curvature of, 146
    curved, 86, 119, 120
        geometry of, 178
        Maxwell's equations in, 86-87
        spinors in, 88-91
    Euclidean, 185-190
        four-dimensional, 185, 186
    flat, 93
        metric tensors of, 90
    functions of, 77, 93
    Minkowskian, 42, 185
    plane symmetry, 106
    points of, 167, 169

Riemannian, 185
V_4, 110, 111
Special relativity, 35-44, 119, 120
    postulates of, 35-37
    principles of, 35
Special unitary group, 6-7
Speed of light, 36, 93, 124, 128, 130, 138
    constancy of, 38
    in vacuum, 41, 99
    invariance of, 44
Spin,
    1/2, 98
        isotopic, 168, 169
    half-integer, 14, 71, 88
    integer, 71, 88
    intrinsic structure, 97-98
    isotopic, 168-169, 172
        conservation and invariance of, 168-169
        conservation law of, 168-169
        orientation of, 169
        rotation of, 169
    total, 168, 172
Spin affine connections, 94
Spin covariant derivative, 93-94
    spin form, 93
    vectorial form, 93
Spinor affine connection, 92-93
Spinor ray, 91
Spinors, 83, 88, 91, 95
    complex conjugate of, 91, 97
    components of, 71
    conformal, 155, 156
    covariant derivatives, 92-95
    curvature, 143-146, 147, 156
        dual to, 147
        symmetry of, 145-146
    Dirac, 176
    Einstein, 181-182

electromagnetic field, 95-99
  decomposition of, 96-97
  electromagnetic potential, 95
η, 180, 181
energy-momentum, 174-176
field strength, 193
four-component, 98, 99, 176, 187
gauge field, 167-184
  Euclidean, 185-195
gravitational, 150-152, 155, 156
gravitational field, 109-165, 180
  trace of, 180
Hermitian, 91, 95
indices of, 88, 89, 176-177
  contraction of, 89
  primed, 89, 91
  raising and lowering, 88, 89
  unprimed, 91
infinite-dimensional, 77-80
isovector equivalent, 176
Levi-Civita metric, 88, 89, 92
Maxwell, 97, 173
mixed-indices, 183
one-index, 144
order of, 70
Pauli, 98
product of, 91, 144
ψ_E, 188
Ricci, 152-153, 181, 182-183
  tracefree, 156, 181, 182
Riemann, 146
set of, 91
SL(2,C), 176, 177, 192
SU(2), 176, 177
tensor equivalent to, 90, 91
tracefree Ricci, 156
transformation law, 77
two-component, 65-82, 83, 88, 93, 98, 190

O(4), 190-192
ξ, 180
Weyl, 152, 153-156
  conformal, 153, 155, 156, 180
Yang-Mills, 172-174, 176, 177, 182
ζ, 181
Standard isomorphism, 60
Subgroup, 1-3
  invariant, 3
  normal, 1, 3
  one-parameter, 73
  of L, 73
  of SL(2,C), 73
Subspace,
  closed, 30, 31, 32
  finite-dimensional, 30
  invariant, 12, 30
  null, 12
Summation convention, 44
Sun,
  gravitational acceleration towards, 119
Superconductors,
  properties of, 133

Tensor density, 86, 112
  weight of, 86, 112
Tensors, 88, 111-112
  conformal, 95
  contraction of, 112
  contravariant, 112
  covariant, 112
  curvature, 98, 143, 144, 145, 146-147, 148-149, 150, 178
  density of, 86
  derivative of, 114
  Einstein, 95, 116, 125
    spinor equivalent to, 153

INDEX

electromagnetic field, 84, 95, 97, 145
  dual to, 84, 85, 97
  spinors equivalent of, 96, 145, 148, 172
energy-momentum, 121, 122-123, 127, 134, 174, 175, 181
  density of, 134
  electromagnetic, 122-123, 174, 175
  gauge field, 174-175
  of Yang-Mills field, 181
  trace of, 123
  zero, 105
four-index, 177-179
indices of, 88, 89
Levi-Civita, 97
  density, 97
  weight of, 97
Maxwell, 86
metric, 89, 112, 116, 120, 124, 139
mixed, 112
O(4),
  indices of, 191-192
  spinor equivalent of, 190
Ricci, 104, 115-116, 140, 142, 178-179
Riemann, 115-116, 118, 142-143, 146, 149, 150, 151
Riemann curvature, 98, 143, 144, 145, 146-147, 148, 149, 150, 178
  decomposition of, 142-143, 148-149, 155
  dual of, 149, 151, 157
  spinor equivalent of, 146, 148, 149, 150, 151, 152, 153, 155

  symmetry of, 150, 151
  spinor equivalent of, 89, 91
  skew-symmetric, 112
  symmetric, 112
$T_{\mu\nu}$,
  spinor equivalent to, 190
  tracefree Ricci, 142, 152, 179
  spinor equivalent to, 153
  Weyl, 116, 155
  Weyl conformal, 116, 142-143, 155-156, 178
  spinor equivalent to, 155-156
Time,
  rate of change, 127
Time reversal, 47
Transformation, 5, 18, 19
  conformal, 156
  coordinates, 110-111, 117, 119, 124, 126, 127
  cosmological, 41-44
    interpretation of, 43
  duality, 189, 194
  $g_{\mu\nu}$, 113
  Galilean, 37-38,
  gauge, 168
    isotopic, 169-171
  identity, 57
  Λ, 39, 40, 55, 56, 57
  linear, 4, 18, 19, 44, 80
  Lorentz, 38-41, 42, 43, 45, 46, 61, 172
    derivation of, 38-41
    determinant of, 45
    four-dimensional, 45
    improper, 45, 46, 47
    inhomogeneous, 46
    inverse, 40, 41
    matrix of, 39, 40, 47
    orthochronous, 46

proper, 45, 46, 47
Mobius, 66, 67
   image of, 67
Translations, 46
   origins of systems, 37
   SU(2),
   time coordinate, 37
Trautman, A., *164*
Two-body problem, 142

Universe,
   age of, 41
   expansion of, 41-43
   homogeneous and isotropic, 41

Variational problem, 116
Veblen, O., *108*, *165*
Vector current density, 85
Vector function,
   continuous, 31
Vector potential, 85
Vectors, 190
   contravatiant, 110-111, 114
   covariant, 111, 114, 115
   curl of, 114
   O(4), 192
     spinor equivalent to, 191, 192
Velocity, 37, 42, 43, 44
   components of, 38
   coordinate system, 38, 41
   light propagation, 38
   receding of galaxies, 41
   relative, 39
Vilenkin, N.Ya., *33*, *63*, *82*

Van der Waerden, B.L., *10*, *82*, 88, *108*, *164*

Wave function,
   derivatives of, 169
   two-components, 169
Weber, J., 132, *165*
Weil, A., *10*
Weinberg, S., *165*
Weyl, H., 14, *82*
Weyl's method, 13, 16
Whitehead, A.B., 128, *164*
Wigner, E.P., *33*
Witten, L., *163*

Yang, C.N., 169, 170, *184*
Yang-Mills field, 181
   geometry of, 177-183
Yang-Mills theory, 167-172
   Euclidean, 190-194

Zero, 2
Zero-eigenvalue modes, 190
Zero energy-momentum tensor, 105